M

Supralittoral Zone

Supralittoral Fringe

3

2

Midlittoral Zone

1

Laminaria (B)

0

Nereocystis (B)

-1

(See page iv for explanation.)

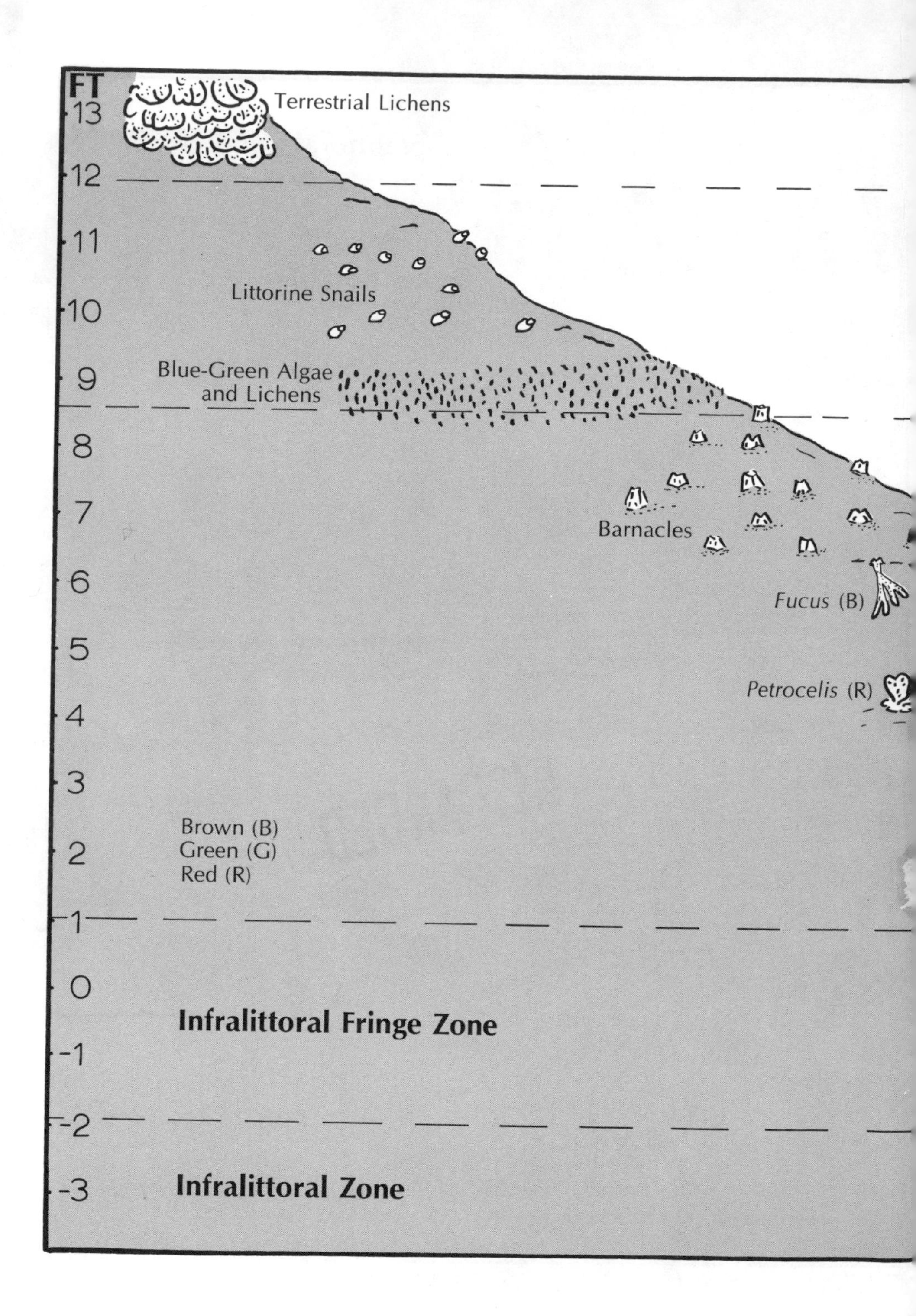
FT
13
12
11
10
9
8
7
6
5
4
3
2
1
0
-1
-2
-3
Terrestrial Lichens
Littorine Snails
Blue-Green Algae
and Lichens
Barnacles
Fucus (B)
Petrocelis (R)
Brown (B)
Green (G)
Red (R)
Infralittoral Fringe Zone
Infralittoral Zone

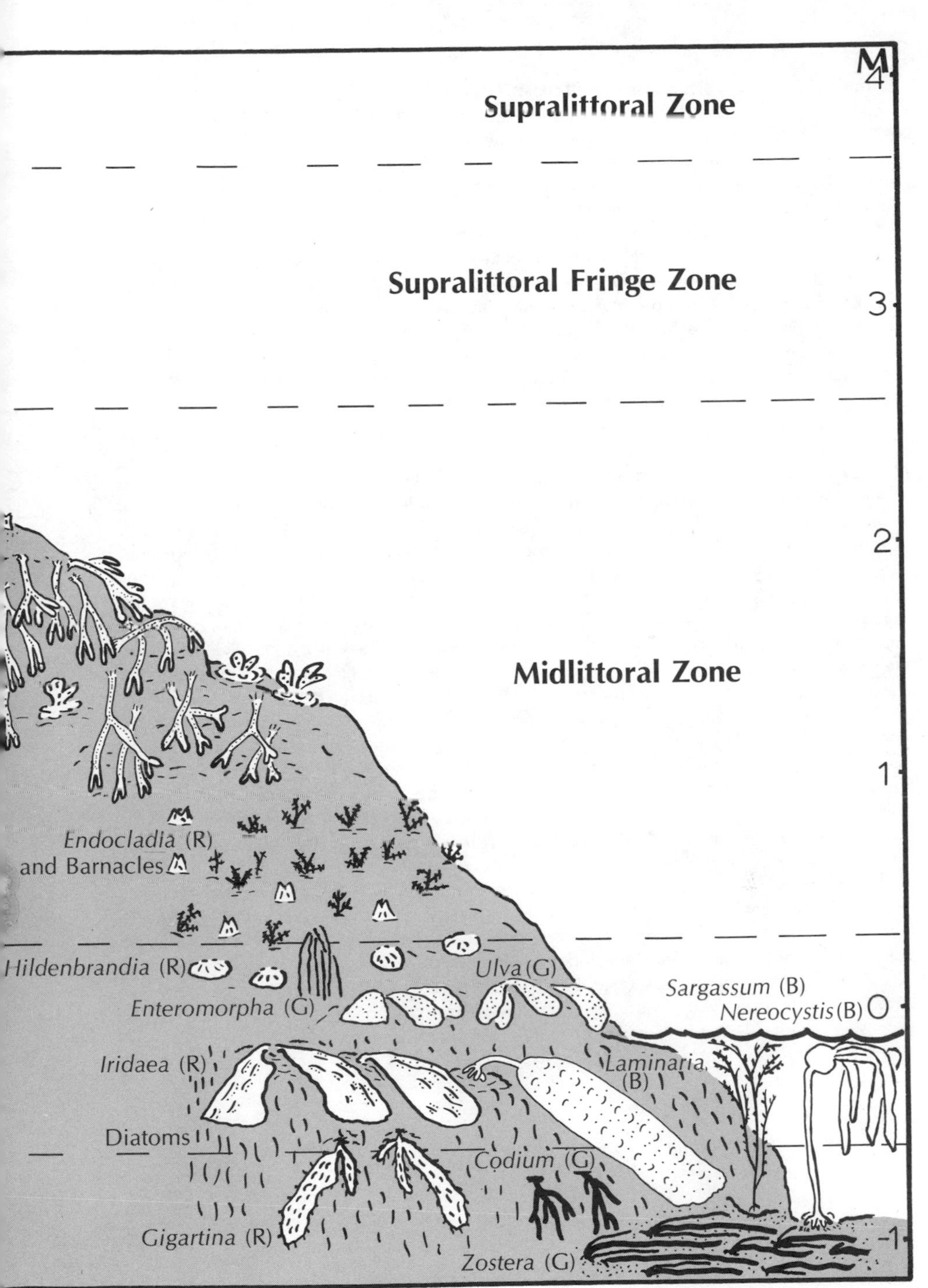

(See page iv for explanation.)

1. Intertidal zonation diagram (exposed coast). Depicted here is Postelsia Point, a fairly exposed location on Waadah Island, Neah Bay, Washington. Between +10 and +12 feet, the green alga, *Prasiola,* and the brown alga, *Ralfsia,* form a distinct zone. At the +9 foot level there is a band of the "wiry" red alga, *Endocladia,* below which the encrusting and upright parts of the red alga, *Petrocelis (Gigartina)* are found. On very exposed parts of the reef, groves of the brown sea palm, *Postelsia,* are usually awash in the surge at the +7 foot level. At about the +6 foot level, groups of the sausage-shaped red alga, *Halosaccion,* are commonly found. From +5 to +2 feet, the feather-shaped blades of *Alaria* droop down over the rocks when the tide is out. Below the *Alaria,* one finds the stout, almost "woody" *Lessoniopsis* plants whose many long, narrow blades are swept to and fro by the surge. *Laminaria* occurs in the very lowest part of the intertidal zone and is submerged on all but the lowest "minus" tides. The permanently submerged infralittoral or sublittoral region often has *Nereocystis* as a prominent member of the shallow water subtidal community, as shown here. (Diagram based on data of Rigg, G. B. and Miller, R. C. *California Academy of Sciences.* Proceedings, San Francisco, 1949. Series 4, vol. 26, no. 10, pp. 323-351.) *(See inside front cover.)*

2. A typical intertidal zonation pattern on a sheltered shore in the inland waters of British Columbia or Washington. This diagram is based on data for Brandon Island near Nanaimo, Vancouver Island. United States tidal datum has been used here; subtract 3.8 feet to convert to Canadian tidal datum. In the supralittoral zone (above +12 feet) one finds a growth of terrestrial lichens. Just below this (+10 to +12 feet), in the supralittoral fringe zone, there are numerous periwinkles (small snails of the genus *Littorina*) but few obvious algae. Near the bottom of the supralittoral fringe zone lies a very conspicuous belt of black vegetation, which consists of a marine lichen, *Verrucaria,* and blue-green algae. Below the black belt, near the top of the midlittoral zone (+7 to +8 feet), barnacles are abundant. Lower (+3 to +6 feet), a broad band of variable width of the fleshy brown rockweed *Fucus* is interspersed with the red alga *Petrocelis* in both its forms—the thick, very red encrusting phase and the phase of papillate upright blades. In the lower midlittoral zone (+1 foot to +3 feet) the tufts of the wiry red alga *Endocladia* in association with the acorn barnacle, *Balanus,* form a conspicuous community on rocky shores all along the Pacific coast. Near the top of the infralittoral fringe zone (+1 foot), one finds the very thin encrusting red alga *Hildenbrandia* and the green sea lettuce, *Ulva* (0 to +1 foot). In brackish pools or seepages, the tubular green alga *Enteromorpha,* shown at this level, is often encountered. In the lower part of the infralittoral fringe zone (0 to −1 foot), a conspicuous turf or "beard" of diatoms *(Navicula* or *Schizonema)* grows on rocks and other algae. The larger algae of this level include the iridescent red alga *Iridaea* and kelps, such as *Laminaria* and *Alaria.* In the very lowest part of the infralittoral fringe zone (−1 foot to −2 feet) and in the permanently submerged infralittoral zone (below −2 feet), one finds more kelps, such as *Nereocystis,* and the large brown seaweed *Sargassum*; red algae, such as the "Turkish towel" *Gigartina*; and sometimes the green alga *Codium.*In areas with a muddy bottom, the flowering plant *Zostera,* or eelgrass, forms extensive meadows. (Diagram based on data of Stephenson and Stephenson, *Life Between Tidemarks on Rocky Shores,* pp. 205-223.) *(See page ii.)*

Common Seaweeds

of the Pacific Coast by J. Robert Waaland

Pacific Search Press/nature

Cover and Page Design by Lou Rivera
Illustrations by Luann Bice
Diagrams by J. Robert Waaland
Cover Photo: *Callophyllis flabellulata*

*Published in the United States of America
by Pacific Search Press, 715 Harrison Street, Seattle, Washington,
and in Canada by J.J. Douglas Ltd.*

*ISBN 0-914718-19-3
Library of Congress Catalog Card Number 76-58250
Manufactured in the United States of America*

Table of Contents

I wish to thank my wife, Dr. Susan Drury Waaland, who accompanied me on most of the field trips necessary to obtain the photographs for this book and who read and criticized the manuscript. Thanks are also due to Dr. Peggy Hudson for her review and criticism of the manuscript.

A Note on the Units of Measurement

Most of the dimensions used in this book have been presented in the metric system, since we are gradually but inevitably switching to the metric system from the English system. Because most of us, the author included, first learned to consider measurements using English system units, it seemed wise to present a scale comparing the two systems and to give some convenient conversion units. In this book, only three units of measurement from the metric system are used:
meter, centimeter, and **millimeter.**

Meter (m): 1 m = 39.37 in. or a little more than 3 ft.
Centimeter (cm): 1 cm = 1/100 m = 0.39 in.
It is convenient to remember that 2.54 cm = 1 in.
and for greater lengths,
that there are approximately 30 cm per ft.
Millimeter (mm): 1 mm = 1/1000 m = 0.039 in. There are 10 mm per cm and 25.4 mm per in.

Introduction

I

On a planet largely covered by salt water, the boundary between sea and land provides a varied habitat for a rich and diverse assemblage of plants—the seaweeds. Familiar to most of us, seaweeds are often seen growing on rocks at the seashore, drifting in coastal waters, or washed up on the beach. This book is intended to acquaint you with some of the more common species found in the marine waters of the Pacific Northwest. In fact, most of the species mentioned in this book are not confined to this region; they can be found from central California to Alaska.

From a biological point of view, seaweeds are macroscopic marine algae. Among them are a few flowering plants, sometimes called sea grasses, which also live in our coastal waters.

The cool temperate waters of the Northwest coast support an abundant seaweed flora rivaled by only a few other areas in the world in number of species and in the luxuriant growth produced by many of the larger varieties. This region provides considerable diversity of habitat, ranging from the rocky stretches of the outer coast which are often exposed to the full fury of the Pacific Ocean, to the waters of Puget Sound, Hood Canal, and the San Juan Islands which represent a relatively quiet and protected refuge.

In addition to their diversity and beauty, seaweeds supply us with a number of useful products. They also provide living space for other organisms among their tangled branches or under their "leafy" canopies. They are food for a number of organisms. In some places, certain large brown seaweeds, the kelps, form extensive undersea forests; in other areas, mats of delicate, filamentous red algae carpet the sea floor. Near the low tide line, one often finds rocks which are pale pink, so colored because they are covered by thin films of heavily calcified red algae. Once thought to be coral animals, these red algae are now known to be plants which envelop the rocks on which they grow. The observer fortunate enough to have a good hand lens (10X) or, better yet, a microscope, will reap even greater rewards in visually exploring seaweeds since many of the most aesthetically rewarding specimens are very small. In fact, careful microscopic examination is often essential for critical identification of certain seaweeds.

Before examining some particular Northwest seaweeds, it is important to understand the relationship of seaweeds and other algae to other kinds of plant life. The examples in this book represent only a small fraction of the many existing species.

Algae

With the exception of a few flowering plants, alluded to above as sea grasses, all the other seaweeds are members of a diverse assemblage of plant groups called algae (singular—alga). These plants possess the green pigment, chlorophyll, which is essential for absorbing light used in the sugar-producing process known as photosynthesis. Some of the seaweeds, *Ulva* or sea lettuce, for example are bright green, due to the presence of two forms of chlorophyll and a general scarcity of other strongly pigmented chemicals. Other seaweeds, however, are noted for their bright red or strong brown colors. In these examples, there are other pigments present, in addition to the green chlorophyll; these act as "antennae" to energize photosynthesis and mask the green chlorophyll. In such plants, the chlorophyll can be extracted and thus detected by treating the plants with alcohol or acetone in which the chlorophyll will dissolve. The masking pigments may be red or blue (as in red algae) or golden or brownish (as in brown algae). The exact color of the plant in nature depends on the kinds, the amount, and the proportion of the different pigments, so that some algae appear very dark red, almost black, while others appear quite yellow. Though the ability to synthesize a particular pigment is genetically determined, the amount of pigment in a particular specimen depends on a number of environmental variables, such as the color and intensity of the light, the depth at which the plant is growing, and the nutrients available in the water. Some examples of these features will be mentioned later in more detail.

In their ability to photosynthesize, algae share a common property with all the other kinds of "green" plants such as mosses, ferns, gymnosperms, and flowering plants. This ability to make their own food by capturing the radiant energy in sunlight is a process that distinguishes chlorophyll-containing plants from plants lacking chlorophyll, such as fungi (mushrooms and other molds) which nourish themselves by digesting organic matter which may be either dead or alive.

In addition to their photosynthetic pigments, algae are distinguished from most green land plants by the absence of special "woody" cells which are part of a water-conducting or vascular system. In green land plants, woody cells (called tracheids because they resemble microscopic windpipes) have specially thickened walls which not only function as water-conducting passages, but also provide mechanical support for the plant body, keeping it upright in the aerial environment of the land. For algae which for the most part live submerged in a watery world, no strong supporting tissues are needed; nor is a water-conducting system necessary for life when the whole plant is bathed in water or splashed by surf during much of its existence.

Thus, in their possession of chlorophyll and in their lack of a sturdy water-conducting tissue system, algae differ from all other kinds of plants—except mosses and liverworts. Here the main distinguishing feature is a structural one, associated with the form of the reproductive organs which in mosses and liverworts are multicellular and surrounded by a jacket of sterile cells. This does not occur in algae. However, since there are no mosses or bryophytes in marine waters, this is purely an academic digression and not a distinction to be concerned with while exploring the seaweeds. There are some plants, called lichens, which are actually composed of an alga and a fungus living together in symbiotic association (see Chapter VI for a definition of symbiosis). Each such association has a definite and distinct form, resembling neither the individual alga nor the individual fungus. A few marine lichens live high in the splash zone at the seashore—some of the colorful yellow and orange ones are highly visible, while the drab gray or black ones may be less visible.

Seaweeds

Two main features separate seaweeds from other algae. These distinctions are rather simple and straightforward: first, seaweeds are algae that live in marine waters (which eliminates all the freshwater algae); and, second, seaweeds are marine algae that are macroscopic in size, readily visible with the unaided eye (which eliminates all microscopic algae, including both the floating, planktonic forms and the bottom-dwelling, benthonic ones). In other words, seaweeds are those small to very large plants you will encounter on your visits to the seashore.

Phylogeny and Evolution of Algae

At the present time, there are many major groups and subgroups of algae. It is difficult to trace precisely their origin and evolutionary relationships. Some deposits of fossil algae are large and conspicuous. These include many tropical coral reefs which are actually deposits of the calcareous parts of fossilized red algae and large deposits of diatomaceous earth which are accumulations of the silica cell walls of billions of microscopic algae.

Microscopic examination of thin polished slices of certain rocks has revealed 1 and 2 billion-year-old fossil algae which resemble modern blue-green algae. More complex algae, recognizable as green and red seaweeds, have been found in 400 million-year-old fossil deposits. While these examples attest to the antiquity of major algal groups, comparatively few kinds of algae are preserved as fossils, so there are major gaps if one tries to use fossils to trace the evolution of modern algal groups. Thus, for the most part, our concepts of evolution and phylogeny among the algae must be based on evidence provided by examination of present-day algae. The evidence used consists of biochemical and structural information.

Within a particular algal group, single-celled forms are considered the most primitive, filamentous forms more complex and more highly evolved, and forms with solid tissue most complex and hence most highly evolved. Other evolutionary pathways are recognized: colonial forms in which nearly identical cells are arranged in a matrix or joined together in a definite pattern; siphonous and multinucleate forms in which cells are usually very large and have many nuclei; and a type known as the false-tissue type in which tightly joined aggregates of filaments are arranged in such a way that the plant appears to be composed of solid tissue.

Figure 3 shows the possible relationships among the major groups of algae. Included in this diagram are branches representing some groups of algae found primarily in fresh waters or which are microscopic in size and thus omitted in this book.

The Cyanophyta (blue-green algae) are believed to be ancestral to all the other types of algae. Blue-green algae have a very simple type of cell structure, indicating that they are most closely related to bacteria. They were certainly the earliest algae to evolve, for fossil blue-green algae 2 billion years old have been found. The Rhodophyta (red algae) are placed near the blue-green algae in this diagram since they have similar photosynthetic pigments, a similar

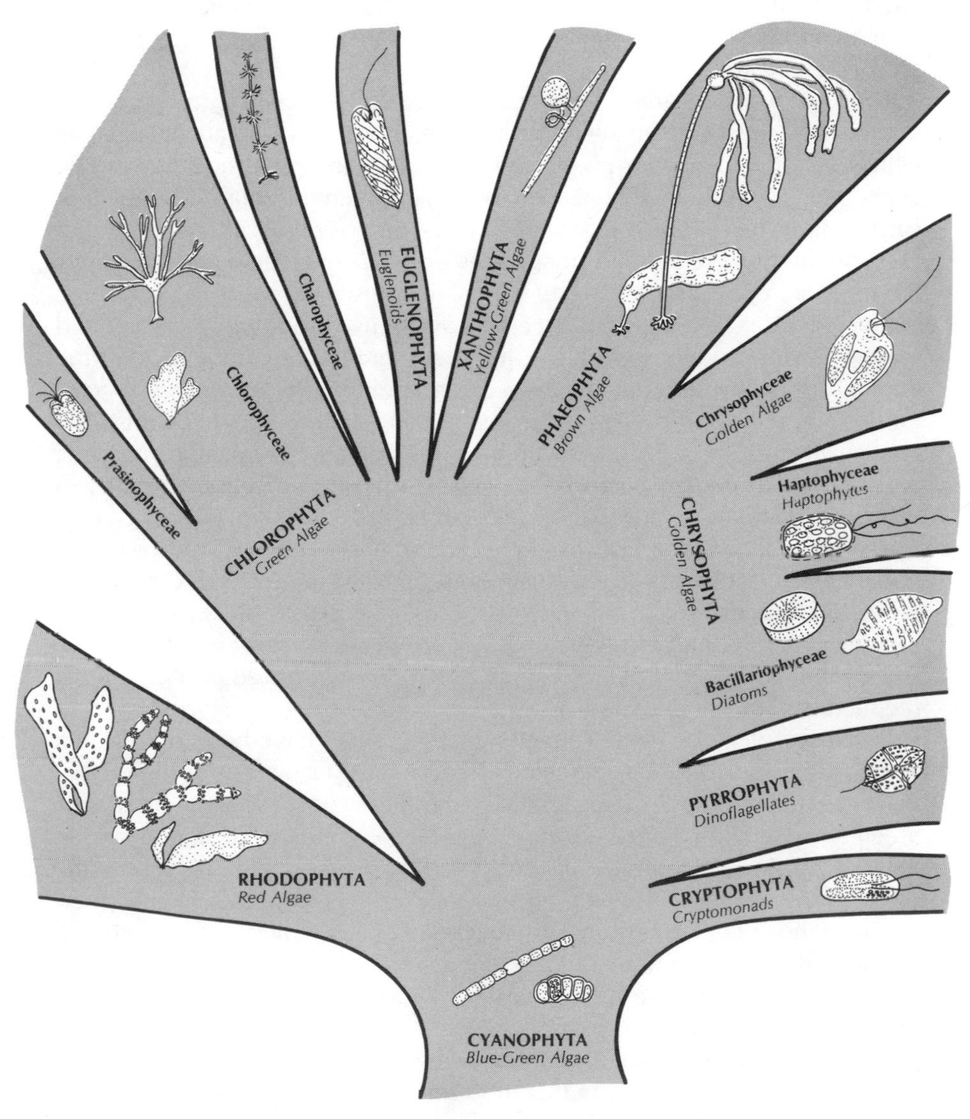

3. Phylogeny of major algal groups. This diagram summarizes the phylogeny and possible evolutionary relationships among the major algal groups including inhabitants of marine and fresh waters.

arrangement of photosynthetic membranes, and no swimming cells. Opposite the Rhodophyta, are the Cryptophyta (cryptomonads) which are planktonic algae with some pigments similar to red algae and blue-green algae, as well as some similar to the other algae on the right-hand side of this chart. Cryptomonads have a distinct cell structure; the arrangement of the pigments in their photosynthetic apparatus differs from that of the red and blue-green algae.

The rest of the diagram is best considered in two major groups, with a number of subgroups. On the upper left side, there is an assemblage of algae which is primarily green pigmented, while on the right side is a major assemblage of primarily golden or brown pigmented algae. These color variations are due to major differences in the composition of the photosynthetic pigments of these algae. Among the assemblage of green pigmented algae are the Chlorophyta (green algae) which are shown here divided into three major subgroups: the Prasinophyceae, a distinctive group of freshwater and marine phytoplankton; the Chlorophyceae, a group containing most of the familiar freshwater and marine green algae; and the Charophyceae, a structurally distinct group of freshwater green algae called the stoneworts. The Euglenophyta (euglenoids) also contain pigmentation typical of "green" algae but have certain distinctions: they are mostly unicellular, possess a distinct cell structure, and are common in fresh waters.

The green algae, and the Chlorophyceae in particular, are believed to be the ancestral stock from which green land plants were derived. Green land plants include mosses and liverworts, ferns and fern allies, gymnosperms including the familiar conifers, and flowering plants. No comparable development of terrestrial plants with a golden, brown, or red photosynthetic pigment system has occurred.

In *figure 3,* the Xanthophyta (yellow-green algae) have been placed in a position between the green pigmented algae and the brown and golden algae. For a long time, some of these algae were thought to be most closely related to green algae, while others were thought to be most closely related to golden algae. Evidence is accumulating which should tip the scale one way or the other in the future, but for the present they are being placed in this intermediate and uncertain position. The major assemblage of brown and golden pigmented algae shown on the right side of the diagram has been called the Chromophyta by some authors, since these algae do share many similarities in pigmentation and other features.

The Phaeophyta (brown algae) are the brown seaweeds discussed in this book. They are noteworthy in that their simplest known representatives are filamentous—already of fairly complex organization. The Chrysophyta (golden algae) consist of three major subgroups: the Chrysophyceae (also called the golden algae), a group which contains many freshwater and marine algae most of which are single celled or, if multicellular, are still of small size; the Haptophyceae (haptophytes), small algae which have a distinct appendage called a haptonema and which may also have distinctive calcareous scales (coccoliths); and, the Bacillariophyceae (diatoms) which have very distinctive cell walls containing silicate. The last group is the Pyrrophyta (dinoflagellates) which have a very unusual nuclear organization and a very characteristic cellular organization, but nevertheless have pigments which indicate a relationship to the golden and brown algae.

Thus, the plants called algae are a diverse group. The macroscopic marine algae—or seaweeds—are found in three main taxonomic groups: Chlorophyta, Phaeophyta, and Rhodophyta.

Seaweed Names

Since the study of marine algae is, for the most part, the province of specialists, only a few varieties have been given common names. Most seaweeds are called by the Latin names set down by the marine botanists who first described the species. Generally, those plants that do have common names are used for food, such as dulse *(Palmaria* [*Rhodymenia*] *palmata),* or are harvested for the chemicals they contain, such as the giant Pacific kelp *(Macrocystis),* or have some particular eye-catching feature, such as the rough surface of the "Turkish towel," *Gigartina exasperata.* The correct Latin names of algae are the only common denominators for clear communication about a particular species and the only realistic way to find out more about a particular alga. The specialized works on algae just do not risk the ambiguity inherent in using common names.

The Latin name of a particular kind or species of alga consists of several parts. There are always two italicized or underlined Latin words: for example, the name of the green alga commonly called sea lettuce is *Ulva lactuca.* The word *Ulva* refers to the genus (plural—genera), of the plant. Genus is a broader category than species; it may include many species, as does the genus *Ulva.* On the other hand, there are many genera—some of our Pacific Coast kelps, for example—which have only one species. The Latin *lactuca* denotes a particular species of the genus *Ulva.* To avoid confusion with other species of algae that might also have the species name *lactuca* as part of their name, one must use the two-word name, or binomial, *Ulva lactuca* (or *U. lactuca,* for subsequent references to the same species in a particular work) when referring to sea lettuce. Another part of the name of a plant is a nonitalicized reference to the author who first described the species. Thus, the full name for sea lettuce is *Ulva lactuca* Linnaeus, since the famous Swedish botanist Carolus Linnaeus was the author who gave this plant its name. Citation of the author's name aids the botanist who wishes to know who first described the species or who first made the binomial combination which the author's name follows.

Scientists who study algae are called algologists or phycologists. (The first term is based on the Latin root; the second is derived from the Greek root referring to algae.) From time to time, algologists restudy certain seaweed groups, and reexamination often results in new interpretations of the relationships between particular algae and their nearest relatives. When this occurs, Latin names may be changed to reflect a new interpretation, and the older name will then become synonomous with the new name. For this reason, you will often find the same plant called by different names in books published at different times. Although more technical works will list synonomous names for reasons of space and simplicity, shorter books usually do not. Often, tracking down the correct name becomes an adventure in botanical detective work, so do not be disheartened if you see what is apparently the same species of seaweed listed under different names in different books. A flora like that of the Pacific Northwest, with at least 500 species of marine algae, always has a few name changes each year.

The formal name is a handy tool for anyone wishing to learn more about a particular alga. Once a particular specimen is identified, its close relatives can be determined. This procedure works for related species in local waters, as well as for species from other parts of the world.

History

The first botanist to seriously collect seaweeds on the Pacific Coast was Archibald Menzies, who visited the Northwest on a fur-trading vessel between 1786 and 1789, and again with Captain George Vancouver in 1791. The specimens he collected were taken to England where they were examined, described, and illustrated by Dawson Turner.

In the early 1800s, other seaweed collections from the Pacific Coast were likewise sent to European marine botanists, such as C. A. Agardh and J. G. Agardh at Lund, Sweden, and W. H. Harvey at Dublin, Ireland. Most of the botanists who examined and described these specimens did not actually collect them, so they had no firsthand knowledge of these seaweeds in the living state. The first resident marine botanist, or phycologist, on the Pacific Coast was W. A. Setchell of the University of California. The many specimens collected by Setchell and his colleague, N. L. Gardner, in the early 1900s, figure prominently in our knowledge of the marine algal flora of the Pacific Coast from Mexico to British Columbia.

Today, there are several major centers where research on the Pacific Coast flora continues: at Stanford University's Hopkins Marine Station; the University of California (Berkeley, Bodega Bay, Santa Barbara, Santa Cruz, and Irvine); the University of Washington (Seattle and Friday Harbor); and the University of British Columbia.

Seaweed Habitats

Even in spring and summer, when seaweeds are most abundant in Northwest waters, the casual visitor to the seashore may have difficulty locating many marine algae. To avoid such disappointments it is best to know in what sort of habitats seaweeds are likely to thrive and how to time your visit to see the greatest variety. To assure yourself of a rewarding visit, you must usually plan your visit to coincide with the low tides. Although one can find numerous seaweeds without regard to the tides—on the sides or bottoms of floating docks and neglected boats, and, of course, with scuba equipment—the "hipboot biologist" will find the greatest diversity when visiting the right shore at the right time. This section will therefore introduce you to some of the main features of tidal phenomena, discuss some of the habitats in which you are most likely to find seaweeds, and mention some more common seaweeds you might expect to find in each type of habitat.

Tidal Phenomena

If you have spent time at the seashore, you are undoubtedly aware of the rhythmic rise and fall of the level of the sea due to the tides. On the Pacific Northwest Coast, there are usually two high tides and two low tides each day. It is also characteristic of this coast that these two highs and lows are of unequal height or amplitude. Such tides are called *mixed semidiurnal tides (fig. 4)*. Other types of tides include *semidiurnal* (two nearly equal high and low tides per day, such as occur on the Atlantic Coast of North America) and *daily* or *diurnal* (one high and one low tide per day, such as occur in the Gulf of Mexico).

The tides have been called "the largest waves in the ocean." The most important factor influencing these waves is the gravitational pull of the moon and the sun. Tides are highest when the earth, sun, and moon are lined up (near full moon and new moon), and their pull is acting in the same direction.

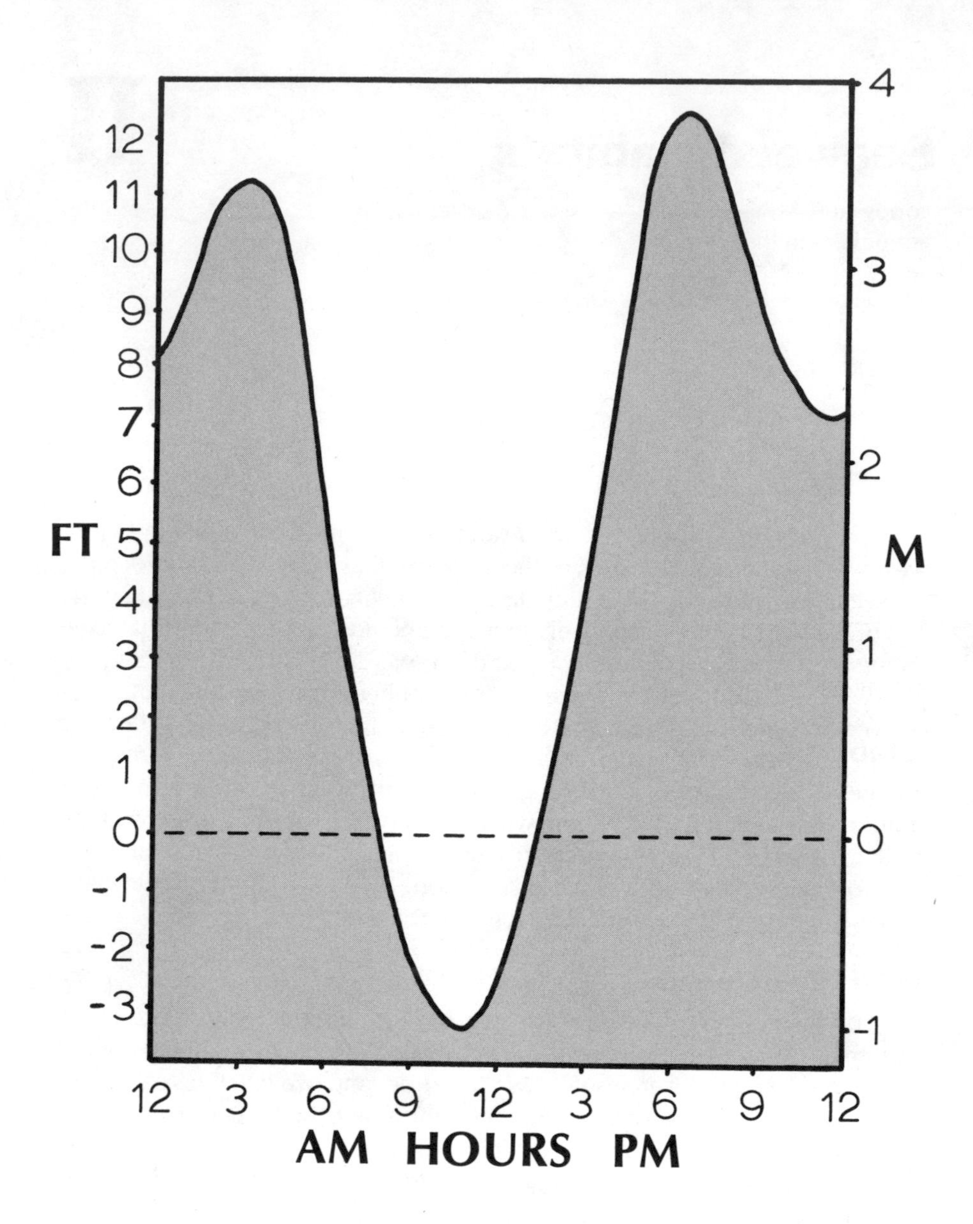

4. A typical tidal cycle during a spring tide on the Pacific Coast where tides are of the mixed semidiurnal type. The shaded area shows the height of the tide during the course of a day. This diagram was based on a Seattle tide in midsummer; on other dates and at other places, the exact height and time of the different high and low tides will vary. (The zero tide level for Canadian tide tables is based on a different zero level than United States tide tables. To convert Canadian tide predictions to the same zero level used here, subtract 2.5 feet from Victoria predictions and 3.8 feet from Vancouver predictions.)

At such times, they are called *spring tides.* When the moon is at right angles to the earth-sun line (near the first and last quarters), the height of the tides is somewhat less, since the gravitational forces are not acting along the same axis. Tides at such times are called *neap tides.* In many parts of the world, the topography of the ocean bottom and the shape of shorelines exert a profound influence on the peculiarities of the tides. Thus the greatest tidal ranges in the Pacific Northwest are found in the upper reaches of Puget Sound, while the total amplitude becomes generally less as one proceeds toward the outer coast.

In your search for seaweeds, you must be aware of the tides, for most of the seaweeds you will see occupy that part of the shore called the *intertidal zone*, i.e., that part of the shore alternately covered and uncovered by the tides. Thus, you should time your visits to the seashore to coincide with the low tides of the day. Tide tables are available at most sporting goods stores, and many newspapers publish the times of the high and low tides. Initially, when your experience with seaweeds is limited, it will be sufficient to go to the shore on nearly any low tide (neap or spring) in daylight hours. Later, when you are familiar with most of the seaweeds in the upper part of the intertidal zone, you will want to time your visits for the lowest of the spring tides, the minus tides, when the intertidal zone is uncovered to the greatest extent.

Stable Substrates

Most seaweeds live attached to some sort of stable substrate or underpinning which is most often rock. A special organ of attachment, called the holdfast, serves to anchor the seaweed to the substrate. Without this holdfast, wave and current action would eventually carry most seaweeds away from the shore and into deeper, lightless waters. Some seaweeds do have inflated hollow portions which keep them afloat; however, most are denser than seawater and would eventually sink into the dark abysses where they would perish from lack of light.

The most typical solid, stable substrate consists of a rocky reef or headland. If you are not familiar with a particular stretch of coastline, you can locate nearshore reefs or rocky areas by consulting a nautical chart or topographic map of the area. *Figure 5* indicates where such sites can be found. Although massive rocky reefs or headlands generally have the most abundant seaweed populations, often beaches with very large boulders—ones not readily moved about by wave action—have moderately extensive seaweed communities. Even such man-made features as rock or concrete breakwaters, concrete ramps, pilings, and seawalls support substantial growths of marine algae and animals. Often, such man-made structures provide the only obvious seaweed communities in areas which otherwise consist mostly of shifting sand, gravel, or mud. On stable features, such as those mentioned above, one can find

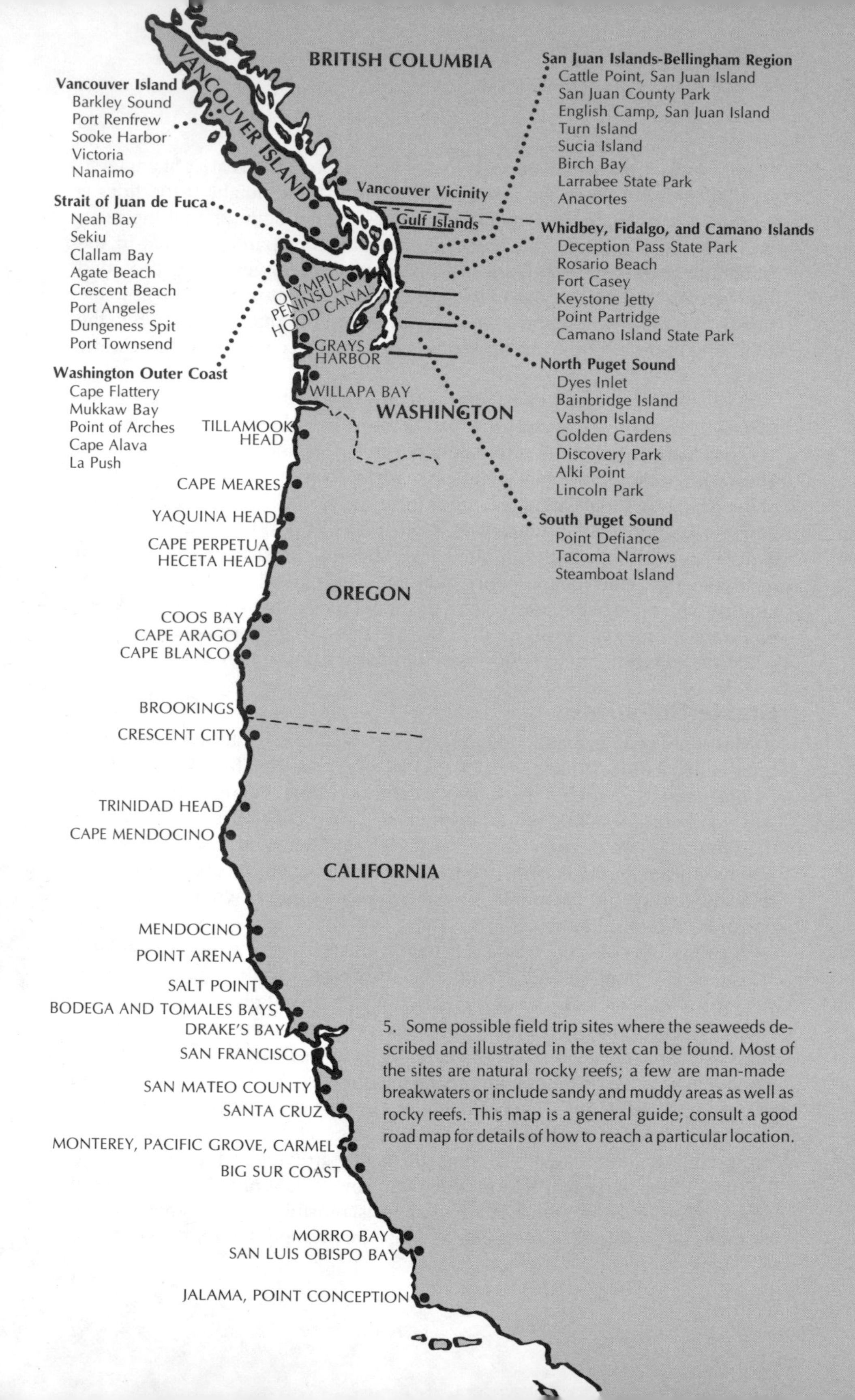

5. Some possible field trip sites where the seaweeds described and illustrated in the text can be found. Most of the sites are natural rocky reefs; a few are man-made breakwaters or include sandy and muddy areas as well as rocky reefs. This map is a general guide; consult a good road map for details of how to reach a particular location.

typical associations of seaweeds—*Fucus, Ulva, Porphyra, Gigartina, Odonthalia, Hedophyllum, Laminaria,* and *Alaria,* to name but a few.

Other stable seaweed habitats include such objects as wooden pilings (where you are likely to find *Bangia* and *Porphyra)* and anchored objects, such as floating docks, barge bottoms, and similar structures where *Polyneura, Polysiphonia,* filamentous diatoms, and some kelps are common. Another habitat—and certainly one not to be overlooked—is the surface of other marine plants and animals. For example, you will only find the red alga, *Smithora,* by examining leaves of the sea grasses, *Zostera* and *Phyllospadix;* one species of *Porphyra, P. nereocystis,* is only found growing on the stipe of the bull kelp, *Nereocystis luetkeana.* Many species of marine algae grow exclusively attached to certain kinds of marine animals. Some of these animals may be sessile; others may move about but rarely or never wander beyond a range in which their hitchhiking seaweeds can survive.

Unstable Substrates

In this category, there are a variety of substrates, ranging from small cobble or shingle beaches to habitats where fine sand, mud, or clay predominate. Seaweeds attached to such substrates can often be readily moved about by wave or current action and one often finds only a few seaweeds in such places. If strong wave action is frequent or severe, the abrasion or scouring due to wave-driven sand precludes the development of extensive seaweed growth. Occasionally, in quiet and still waters, such substrates will support moderate seaweed populations. Often, in sandy or muddy areas one finds scattered stands of seaweeds, such as *Neoagardhiella baileyi* or *Gracilaria sjoestedtii,* occupying bits of solid material, such as small pebbles and shell fragments. Such communities are particularly common in many parts of Puget Sound and Hood Canal. They can also be found immediately adjacent to rocky reefs which support more striking seaweed populations.

More a graveyard than a habitat of seaweeds is the beach located adjacent to subtidal rocky reefs or submerged ledges, strewn with the sort of litter that appeals to the beachcomber. In the days before marine botanists took up scuba diving, the only means for collecting many deep-dwelling marine algae was by use of dredging equipment operated from a boat or by searching carefully through the beach drift. Thus, the serious seaweed student should not overlook the treasures cast up on the beach. Collecting at such sites is often especially rewarding shortly after severe storms.

Eelgrass *(Zostera marina)* forms extensive meadows in many mud-bottomed areas in Puget Sound and Hood Canal. The extensive root and rhizome systems of these flowering plants anchor them firmly in the mud and sand. Bright green strap-shaped leaves arise in profusion from the rhizomes, providing a habitat for other plants and animals.

Seaweed Distribution

If you are successful in locating areas where seaweeds are reasonably abundant, you may observe that seaweeds (and other marine organisms, such as snails, mussels, and barnacles) often grow in distinct bands or zones. If you could peer beneath the surface of the waters, you would find a definite pattern or zonation among subtidal seaweeds as well. In fact, there appear to be several major and many minor patterns of seaweed distribution, with a number of factors important in influencing these distributional patterns. It is easiest to consider seaweed distribution by dividing distributional patterns into the following major categories: intertidal distribution (between the high and low tidemarks); subtidal distribution (below the lowest tide levels); geographical distribution (over long distances); and seasonal distribution (temporal patterns related to seasonal phenomena). Other important and readily discernible factors are exposed versus sheltered habitats and shady versus sunny habitats.

Intertidal Zonation

A rocky shore—rather than a sandy or muddy beach—is the best frame of reference for finding extensive growths of seaweeds. One of the earliest and most frequent activities of marine biologists has been to describe the arrangement of plants and animals on these rocky shores. The arrangement is usually referred to as the *zonation*, since many shore plants and animals occupy distinct zones or bands between the tide marks. Numerous schemes have been devised to name or describe these zones. Some refer to the tidal height (which varies at different places along a coastline and from one coastline to the next); some refer to the principal or conspicuous plant and animal associations found there (the components of these associations may vary from one place to the next); some assign arbitrary numerical designations to the zones; and some refer to the relative tidal heights of various zones. This last practice will be used here, since it has been applied widely to zonation patterns observed on many coasts of the world.

Devised by T. A. and Anne Stephenson after years of observing intertidal life in many parts of the world, this method of identifying zones is relatively simple. It is suitable for recording "widespread" features of zonation and recognizes three main zones within the intertidal or littoral zone. The Stephensons have also assigned names to the regions immediately above and below the tide limits for a total of five named zones. *Figures 1* and *2* (inside front cover and page ii) illustrate the relationships of these zones to one another, and to some of the principal biological and tidal features which delimit the zones. The actual zonation pattern that you will encounter on a particular shore depends on a variety of factors, and the zonation will appear different at different sites. On the Pacific Coast, it is believed that the time or amount of exposure or emersion (time when the tide is out, and the plants are exposed to the air) is the single most important factor controlling the observed zonation. It is also likely that a number of other factors are quite important, but there are so many variables involved that it has been difficult for scientists to sort out the relative importance of each.

Subtidal Zonation

Below the low tide line, different factors come into play *(fig. 36,* inside back cover). Here, seaweeds are never exposed to the air for any length of time. The major factors controlling the vertical distribution of seaweeds on this permanently submerged part of the shore are the amount of light present and the color of the light. Both of these factors vary with increasing depth, and both appear to have a significant effect on the depth at which different seaweeds can survive.

Just below the low water mark, one typically encounters an abundant forest of large kelps *(Laminaria, Agarum, Nereocystis, Pterygophora,* and others, depending on the particular location). In the understory of the kelp forests and in waters deeper than the limits of the kelp forest, there are usually many red algae of medium to small size (30 cm or less). These bright red seaweeds are able to carry on active photosynthesis in dimmer light and in "greener" light than the green or brown algae typically found in shallower waters. Below the limit of these upright algae forms, reside the deepest living algae found in our waters—encrusting coralline algae, which grow as pink crusts on the surfaces of rocks. At greater depths, algae simply cannot grow; there is not enough light to energize photosynthesis. In the inland waters of Washington, the deepest algae are found at depths of about 50 to 60 m and at slightly greater depths where the water is more transparent. In parts of the world where the waters are exceptionally transparent, such as the tropics, algae have been reported growing at depths of 200 m.

Geographical Distribution

While it has been difficult for biologists to pinpoint the most important elements controlling geographical distribution of seaweeds, the single most important factor appears to be the temperature of the sea. In examining the seaweed distribution along a particular stretch of coastline, one occasionally encounters a rather abrupt change in the species composition of the marine flora at certain places. On the Pacific Coast, such a change occurs at Point Conception in California where a cold current from the north meets a warm current from the south. On the Atlantic Coast, such a change is seen on different sides of Cape Cod, where there are significant differences in the sea's temperature. Along the Pacific Northwest Coast, the temperature changes are more gradual; consequently, there are no dramatic examples of abrupt changes in seaweed distribution due to ocean temperature.

Seasonal Distribution

Spring and summer are the best times for finding seaweeds. This is especially true in the Puget Sound region, where a pronounced seasonal change in marine algae can be observed. A few of the larger more noticeable seaweeds are perennials, lasting throughout the winter. Very large, sturdy species, such as *Pterygophora* or *Lessoniopsis* (only found on the outer coast), fall into the perennial category, but even these may lose most of their blades during winter storms. Others, such as *Nereocystis* and *Postelsia,* are annuals which overwinter in a microscopic phase. Their large stage only begins to grow rapidly during the spring; late in the fall they usually succumb to the violence of winter storms. A few seaweeds, like *Iridaea cordata,* have a perennial, crustose base which sends up a large upright blade. This blade grows rapidly during spring and summer, weakens in late summer, and is finally wrenched from its base by fall storms.

The most important factors influencing such seasonal phenomena in Northwest seaweeds appear to be the difference in the amount of light available during winter and summer, and probably the difference in water temperature between the winter and summer months.

Exposed versus Sheltered Habitats

There are major differences in marine organisms found on shores *exposed* to the full force of large waves *(fig. 6),* and those found on shores *sheltered*

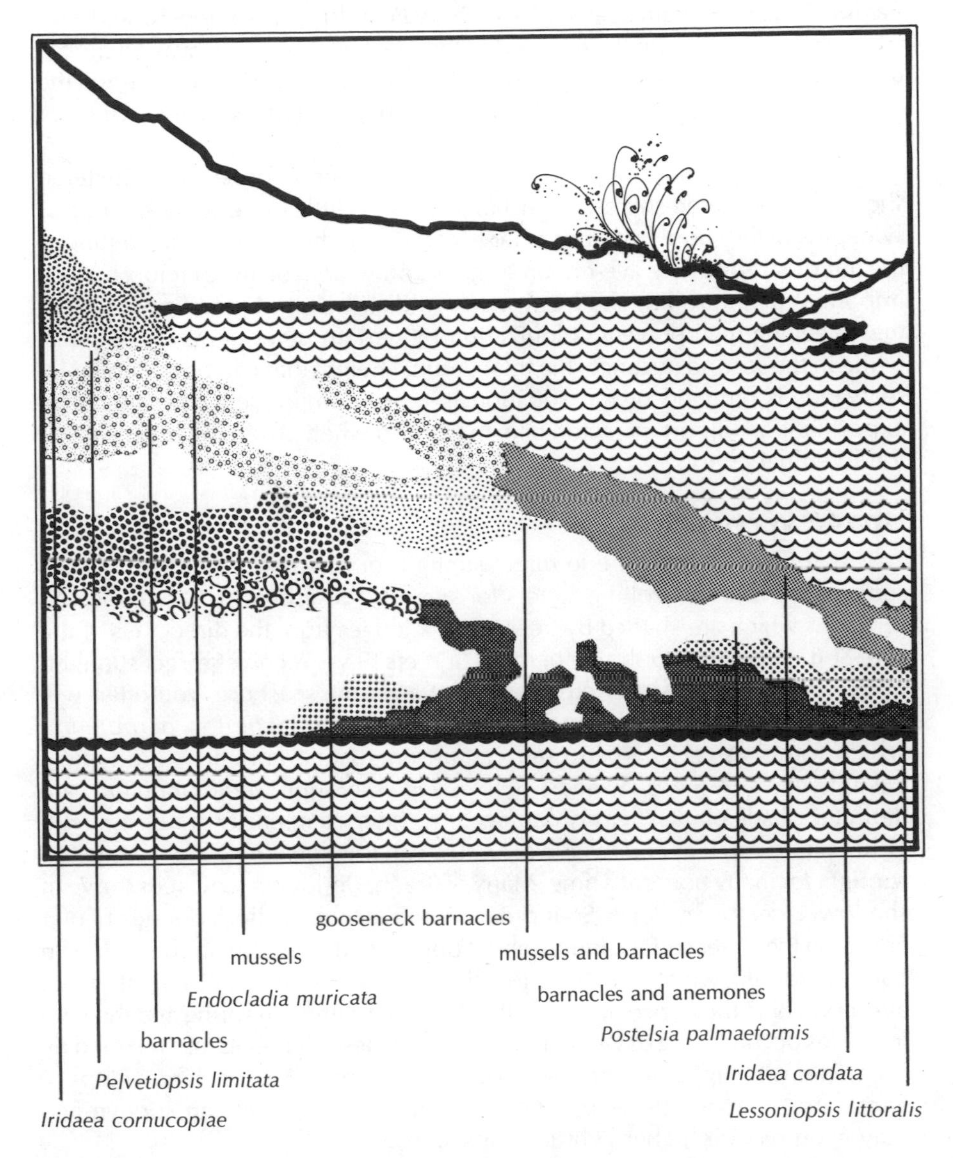

6. Rocky reef - Cape Flattery. Rocky intertidal region on an exposed outer coast. Even when the tide is low, wave action may keep the intertidal zone wet.

from large waves and splashing spray. Most of the inland marine waters of Washington (Puget Sound, Hood Canal, and the San Juan Archipelago) are quite sheltered by comparison with the outer coasts of California, Oregon, Washington, and Vancouver Island which are exposed to the full fury of the Pacific Ocean. Certain seaweeds, such as *Postelsia, Lessoniopsis,* and *Pelvetiopsis,* occur only on the exposed outer coast where violent wave action is common. On the other hand, very sheltered bays and lagoons often permit the development of great masses of free-floating plants of *Monostroma, Ulva,* or *Sargassum.*

There are a number of factors differing between exposed and sheltered shores. On sheltered shores, for instance, there is little or no wave-energized sweeping of the shore by fronds of large algae, and there may be no wetting of the intertidal zone by wave-driven surge or spray. Both of these factors may be important in controlling the local seaweed distribution. In the first instance, the sweeping may prevent colonization by certain other plants or animals, while in the second instance, the total vertical height that can be occupied by seaweeds is reduced or the zonation is more compressed. The degree of exposure is an important factor to keep in mind when observing organisms in the intertidal region.

Shade versus Sun

The amount of exposure to direct sunlight or the degree of shading from sunlight are also controlling factors for seaweed distribution. Often one can find sites which are shaded by trees or rock ledges from the direct rays of the sun. Sometimes the north side of rocks or reefs never receives direct sunlight. In exploring narrow shady crevices or caves at the seashore, you often will find different seaweeds living in such places—*Rhodochorton purpureum,* which forms velvety carpets, and certain encrusting corallines are typical cave dwellers. A few seaweeds, such as *Delesseria* or *Membranoptera,* are definite "shade plants," easily damaged by the high intensity of full sunlight. Other plants, such as *Endocladia* or *Fucus,* can tolerate direct exposure to full sunlight for many hours at a time. Many of the shade-loving seaweeds thrive in shallower depths in Puget Sound during winter when the amount of solar energy at the water surface is only about one-sixth that of the summer value. In addition, during winter the really low tides only occur in the dim light of late afternoon or in the dark of night. As the days get brighter in spring and the very low tides occur during daylight hours, some of these shade plants are killed by high intensity light and/or overgrown by plants like the large kelps—*Laminaria, Nereocystis, Alaria, Costaria,* and others—that can survive and may even require higher light intensities.

Common Seaweeds IV

This section is intended to acquaint you with some of the terminology applicable to seaweeds, general aspects of their overall form and certain specialized structures, some aspects of their life histories, and the sorts of features used in classifying them. Some of the more common seaweeds you will likely encounter in trips to the seashore will also be described and illustrated, as well as their habitat and geographical distribution. Photographs and drawings will help you identify and recognize many of the seaweeds. Bear in mind, however, that there are many more species in Northwest waters than can be illustrated here. Moreover, many seaweeds are quite variable in form; some are very difficult even for specialists to identify. Some will vary in size, shape, or color depending on their age, reproductive state, or location.

Basic Terminology and Seaweed Structure

Like any branch of science, marine botany has developed a plethora of specialized terms which can communicate a great deal of information in just a few words. However, such terms are of little use to nonspecialists and often tend to cause confusion or despair. For these reasons, I have tried to keep the seaweed terminology as simple as possible. There are, however, a few terms related to seaweed structure that are essential if we are to understand these plants. In describing their general form, it is most convenient to start at the base where the seaweed attaches to the substrate.

Holdfast. This is the organ of attachment by which a plant is anchored to a surface. *Plate 1* shows the large and conspicuous holdfast of a kelp, *Egregia menziesii,* consisting of numerous fingerlike projections called *haptera* (singular—*hapteron*). Holdfasts come in a variety of other shapes and sizes. For example, the kelp, *Cymathere triplicata,* has a disc-shaped holdfast resembling a suction cup. In many algae, the holdfast may be very small—only a single specialized cell or a few cells, as in the filamentous green alga, *Urospora*—and you may have difficulty locating it.

Stipe. This is the stalk of the plant. *Plate 1* shows the sturdy, branched stipe arising from the holdfast of *Egregia.* Stipes may be quite long (20 m), as in *Nereocystis (pl. 7),* or short, as in *Cymathere* (5 to 10 cm), or even nonexistent, as in *Hedophyllum (pl. 7).* They may vary from the stiff "woody" stipes of *Pterygophora (pl. 9)* to the flexible stipe of *Nereocystis (pl. 7)* or *Macrocystis (pl. 8).* Some stipes are solid *(Lessoniopsis, Pterygophora,* or *Laminaria),* while others are hollow and filled with gas *(Nereocystis, Postelsia).* It is the function of the stipe to support and orient the blades.

Blade. These are the "leaflike" portions of the plant. There may be only one broad blade *(Ulva, pl. 2,* or *Porphyra, fig. 23),* or the blade may be divided into many smaller parts *(Hedophyllum, pl. 7,* or *Postelsia, pl. 8),* or there may be several blades on a plant *(Macrocystis, pl. 8).* The blade's primary function is to serve as a broad structure for the absorption of light and nutrients in the food-producing process of photosynthesis. Blades may also function as the supporting structures for reproductive structures, such as spore- or gamete-producing cells. Some species of algae *(Alaria, Macrocystis)* even have blades which are specialized for bearing reproductive structures; these special blades are completely covered by sporangia which release millions of spores.

Floats. Especially among the large kelps, such as *Nereocystis* and *Macrocystis,* you will find portions of the plant that are hollow, gas-filled chambers called floats. These keep the photosynthetic portions of the plant buoyant and thus oriented toward sufficient light for photosynthesis.

Thallus (plural—*thalli).* This is a term which refers to the whole plant body of simple plants, such as algae, fungi, and mosses.

Filamentous. Many algae *(Urospora, pl. 3; Ectocarpus, fig. 14;* and *Antithamnion, pl. 13)* have a rather simple type of organization in which cells are joined end to end. These algae are said to have a filamentous organization; some filamentous plants are branched *(Acrosiphonia, pl. 3; Ectocarpus, fig. 14;* and *Antithamnion, pl. 13),* while others are not *(Urospora, pl. 3).* The thalli of many red algae *(Gigartina, fig. 29,* and *Iridaea, pl. 12)* are actually composed of numerous intertwined filaments.

Reproduction

The details of seaweed reproduction and its life cycles or life histories are often quite complex. In essence, most seaweeds reproduce both sexually and asexually. Usually, there is an alternation between these two types of reproduction: First, sexual reproduction occurs by the mating of male and female reproductive cells, or gametes. The zygote resulting from this sexual fusion then germinates to produce a new plant which will usually reproduce asexually by means of spores. These spores then germinate and make gamete-producing plants. *Figures 7, 8, 9,* and *10* illustrate the general patterns of the life histories of four different seaweeds. One of these *(fig. 7)* describes an alternation of isomorphic generations (a type in which the sexual and asexual plants appear similar or identical in overall structure); a second *(fig. 8)* with an alternation of heteromorphic generations (a type in which the sexual and asexual stages are quite dissimilar in size and form); a third *(fig. 9)* where the alternation has been short-circuited or reduced to a few cells; and finally *(fig. 10),* a typical red algal life history where there are actually two spore-producing phases.

Sometimes, a gamete- or spore-producing plant phase is encountered all year round, while in other cases, one phase may be only seasonal. When the life history includes an alternation of heteromorphic generations, you may not recognize the relationship between the two plants, or it may only be possible to find the alternate phase by carrying out laboratory culture studies which require considerable skill and perseverance.

Cell Structure

The knowledge that all living things are composed of small building blocks, or units called cells, was one of the most exciting biological discoveries of the nineteenth century. It is now known that most cells are microscopic in size, and that nothing smaller than a single cell can truly be called alive. The smallest algae are in fact single cells able to carry out all the functions of living things (e.g., *Platymonas,* a small green alga frequently found in high tide pools, *pl. 1*). Although most of the plankton algae (free-floating or swimming algae) consist of only single cells, the cells of some algae (especially many from the tropics) are the largest known plant cells, reaching the size of small hen's eggs. In local waters, the *Halicystis* stage of *Derbesia* is a single cell which may measure 1 cm or more in diameter. It resembles a green pearl *(pl. 4)* and is easily visible to the naked eye, if you know where to look. Most macroscopic algae, however, are complex assemblages of millions of cells (e.g., *Nereocystis).* Often, details of cellular structure differ among algae, and these microscopic features can be of use in identifying them. Different cell types often have different functions: some may be specialized for support, others for photosynthesis, and still others for reproduction.

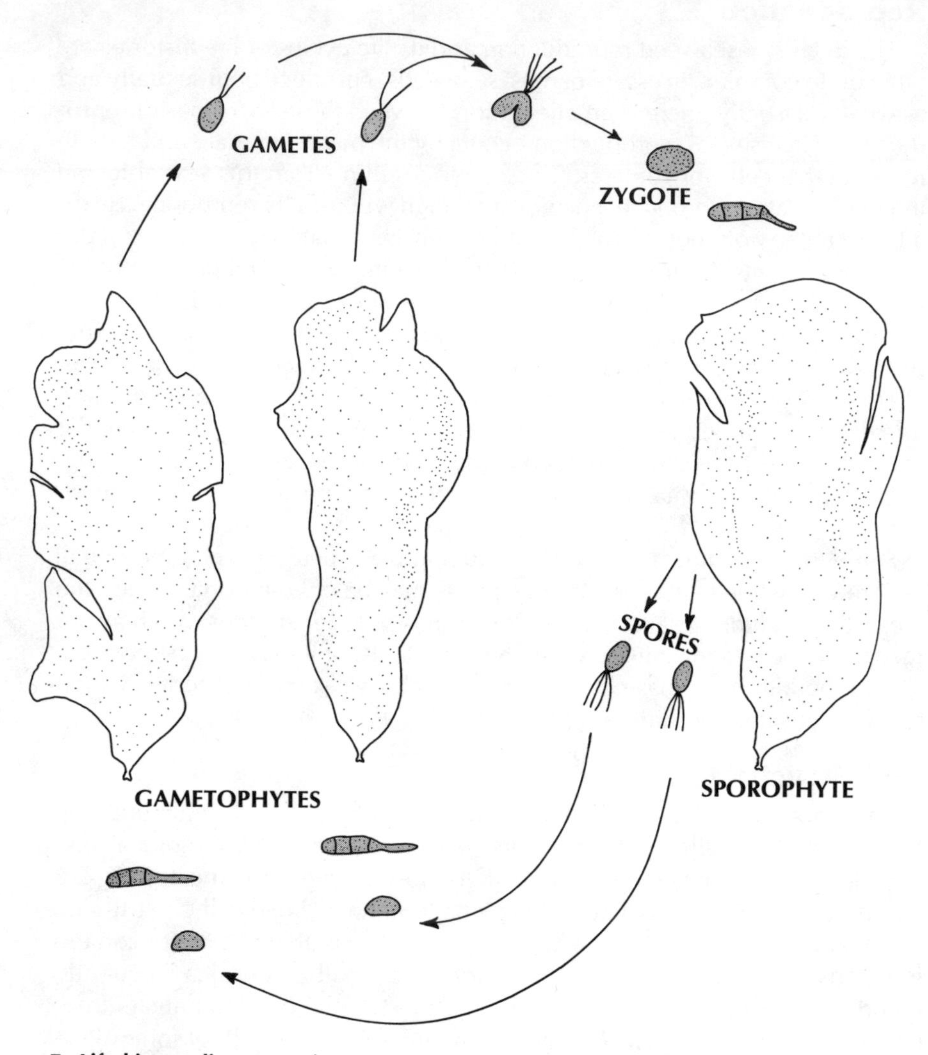

7. Life history diagram - *Ulva.* A green seaweed with an alternation of isomorphic (similar) generations, the macroscopic, blade-shaped gametophytes (sexual plants) produce microscopic, swimming gametes propelled by two flagella. When gametes of opposite mating types meet, they fuse and a zygote is formed. The zygote germinates and eventually grows to produce a macroscopic, blade-shaped sporophyte (asexual plant). When the sporophyte matures and reproduces, it releases microscopic, swimming spores propelled by four flagella; these spores germinate to produce gametophytes. This pattern of alternation between similar-appearing generations should be compared with the other examples presented here to gain an appreciation of the variety of reproductive patterns found in seaweeds.

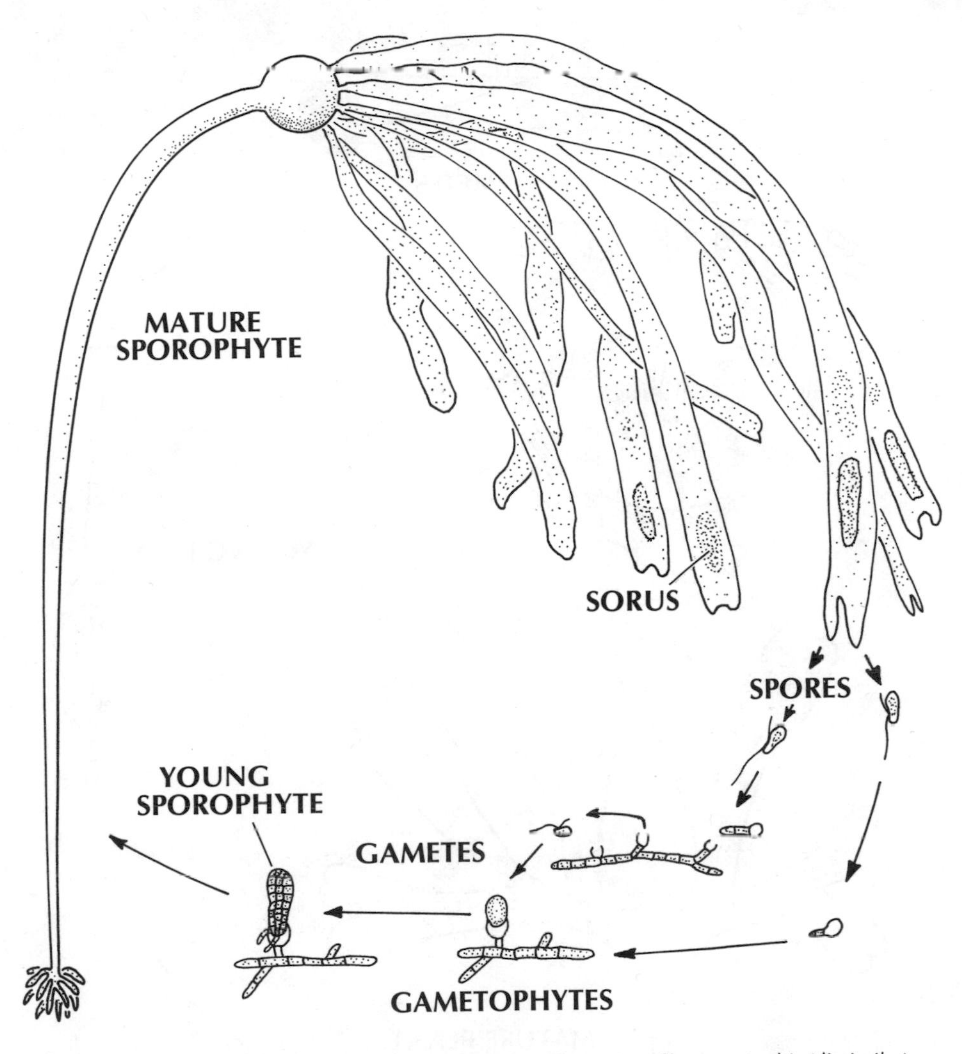

8. Life history diagram - *Nereocystis*. A seaweed with an alternation of heteromorphic (dissimilar) generations, the most conspicuous phase of this species is the sporophyte (asexual) phase. The sporophytes mature in summer, at which time elongate patches (sori) of chocolate brown sporangia form near the ends of the long blades. Entire groups of sporangia drop from the blades and then release microscopic swimming spores which have two flagella. After settling, these spores germinate into microscopic, filamentous gametophytes (sexual) of two different sizes: the smaller ones produce male gametes (sperm), the larger ones produce female gametes (eggs). Sexual reproduction occurs when the male gametes swim to the eggs and fertilize them. Following fertilization, young sporophytes develop from the zygotes, and the large, conspicuous sporophyte develops. The tiny inconspicuous gametophytes of the kelps were only discovered by means of laboratory culture studies starting with the spores. This type of life history permits the seaweed to pass part of the year as a very small inconspicuous plant.

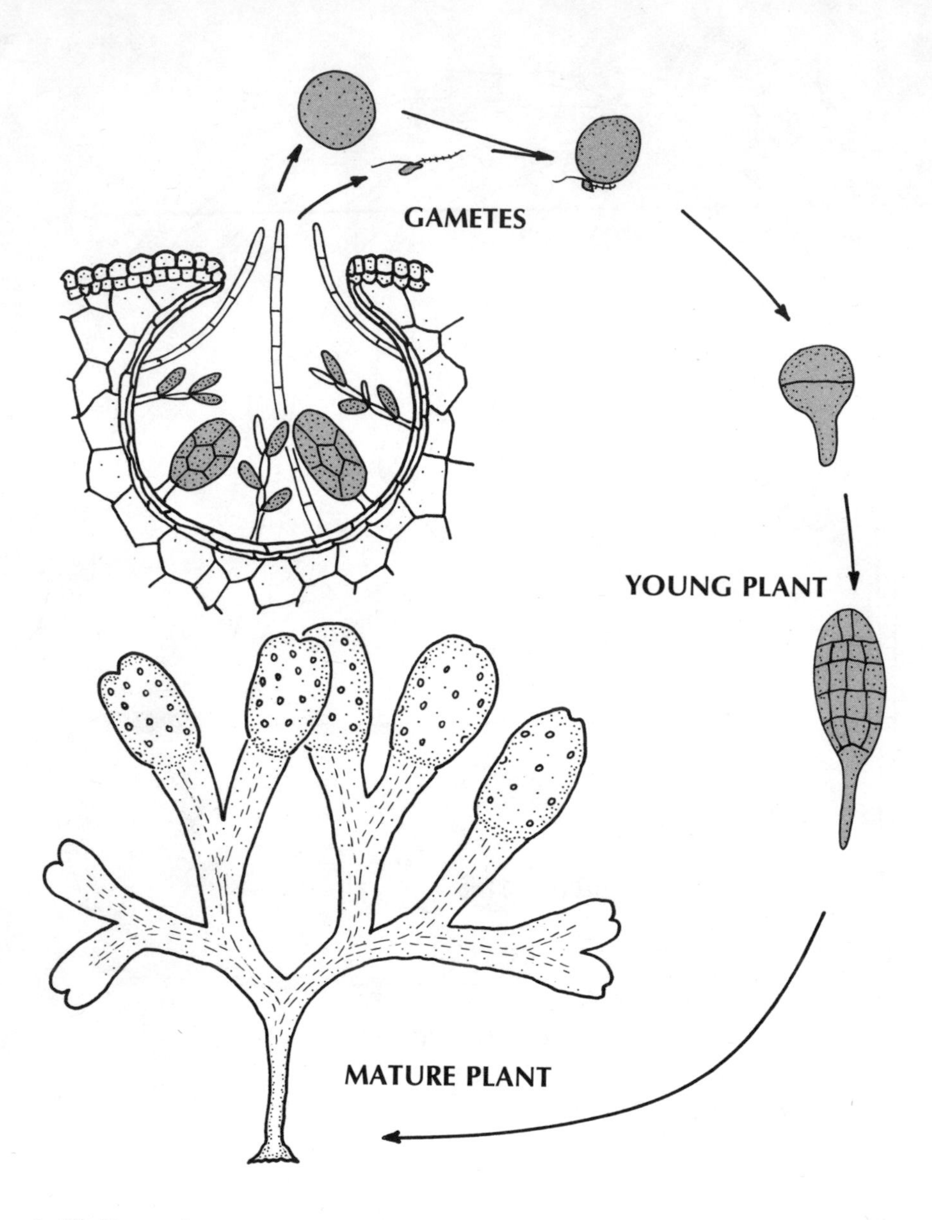

9. Life history diagram - *Fucus.* A seaweed which does not produce a free-living, independent haploid phase, *Fucus* has special organs borne inside minute depressions (conceptacles) on its branches. Here, reduction division, or meiosis, takes place. Gametes are produced following meiosis and released into the water where they fuse and produce zygotes. The zygotes germinate and grow into new *Fucus* plants.

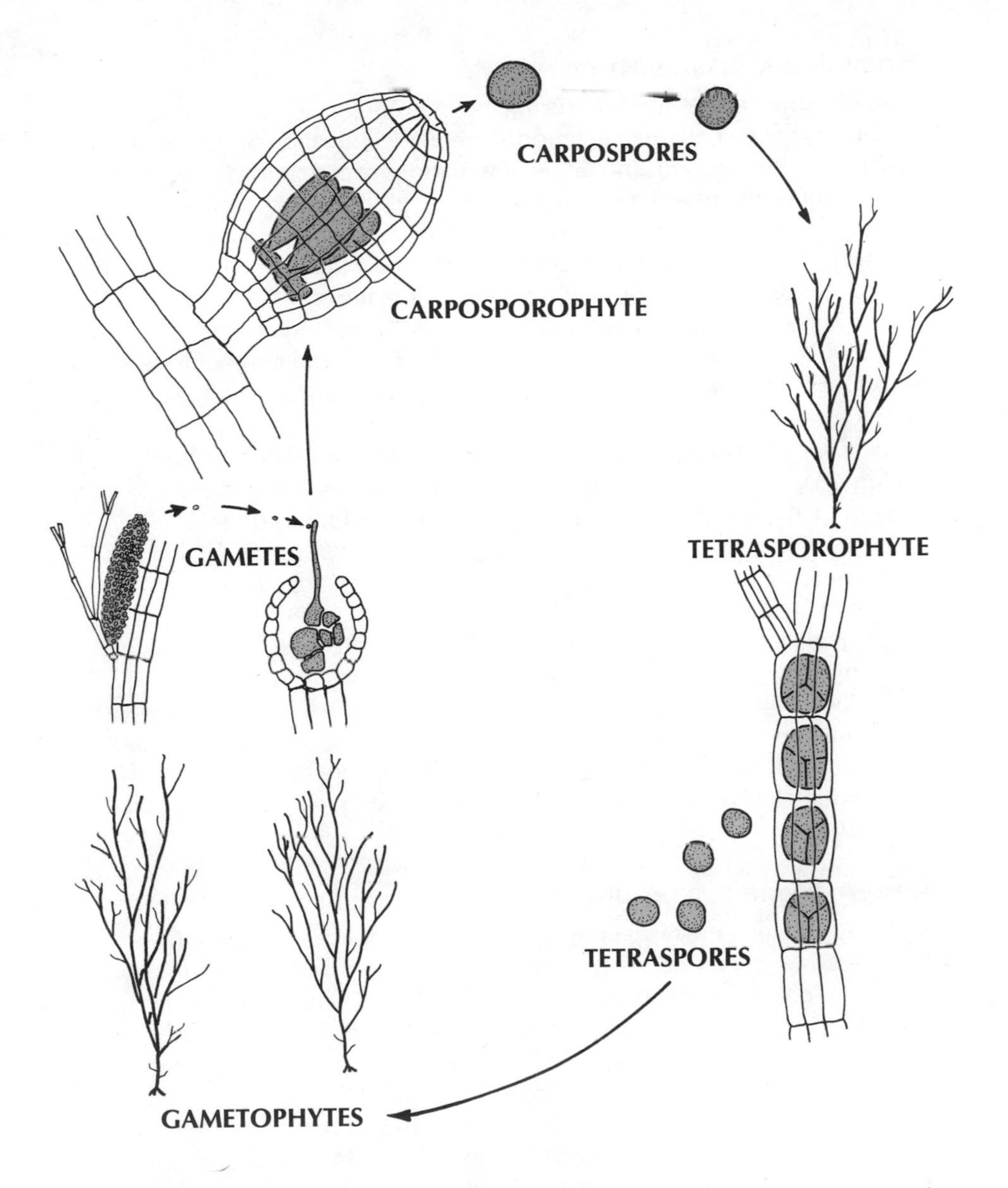

10. Life history diagram - *Polysiphonia*. Following fertilization in sexual plants of many red algae, a tiny, spore-producing phase is produced which is genetically distinct from its parent plants but remains attached to the female plant. This phase is called a carposporophyte, producing many spores called carpospores which then germinate into a free-living asexual phase called the tetrasporophyte. The tetrasporophyte may resemble the sexual plants in its overall form (isomorphic), as in *Polysiphonia,* illustrated here, or it may be different (heteromorphic), as in some other species of red algae.

Seaweed Classification

In the early nineteenth century, the major groups of algae were recognized on the basis of color or pigmentation. Careful identification of the pigments in algae shows that certain kinds and combinations of pigments characterize the major groups of algae. Other characteristics used in determining the major group to which an alga is related include the type of stored photosynthetic product (starch or oil), cell wall composition, and details of cellular structure, plant form, growth pattern, and reproductive features.

The major groups of seaweeds are the green algae (Chlorophyta), the brown algae (Phaeophyta), and the red algae (Rhodophyta). Their probable relationship to some of the other major groups of algae (many of which are microscopic phytoplankton) is shown in *figure 3*.

Within a particular group of algae, such as the brown algae, further classification is determined by the general shape of the plants, the pattern of their growth, and the pattern of their reproduction and life histories.

In providing examples of the more common seaweeds of the Pacific Northwest, we will concentrate on the green, brown, and red algae, since they are the really conspicuous plants of our marine flora. A few examples of other types of algae and marine plants will be included to give you an inkling of plant diversity. I have tried to concentrate on large, easy-to-find plants, but have also included a few algae that are hard to find without an experienced guide to lead you to these prizes. Finally, some of you will most certainly come upon reasonably common seaweeds that are not included in this book. For more information about them, explore the books listed in Selected References.

Green Algae (Chlorophyta)

Because of their bright green color (due to chlorophylls *a* and *b*), their starch food reserves, cellulosic walls, and the structure of their cells, this group of algae is closely related to the green land plants. There are probably more species of green algae in fresh waters than in marine waters; however, those present in marine waters, such as *Ulva* and *Enteromorpha*, are often quite conspicuous. As for genera, there are many found only in marine waters, and more of these types of green algae are found in warm, tropical seas than in cool, temperate waters, like those of the Pacific Northwest. Nevertheless, there is a fair cross section of green algae to be found here—more than thirty genera with perhaps ninety species. A number of these are discussed and illustrated in the pages that follow.

Monostroma arcticum Wittrock *(Pl. 2)*

Description: *Monostroma* forms green sheets or blades (10 cm by 15 cm long)

which are only one cell thick. There are five species of *Monostroma* in Northwest waters; they may superficially resemble *Ulva* which differs in being two cells thick. *Monostroma* begins life as a small green cushion of cells which develops into a small sac. As the plant grows, the small sac splits and each segment resulting from the split grows into a large blade. The blades may remain attached to the cushion-shaped holdfast or may break off and continue to grow if the fragments remain in appropriate waters.
Habitat and Distribution: *Monostroma* is reasonably common intertidally. It grows attached to small stones, as well as to solid reefs. In sheltered, often brackish bays, free-floating sheets of *Monostroma* may persist for some time, reaching 1 to 2 m in size. One species, *M. zostericola,* occurs as small, pale, thin green blades growing on the surface of eelgrass or surfgrass. Many species of *Monostroma* are quite cosmopolitan: along the Pacific Coast, *M. arcticum* ranges from Alaska to Washington. There are also tropical species of this genus.
Comments: In the Orient, *Monostroma* is often used as a condiment and has been cultivated in much the same manner as *Porphyra* (see Chapter V).

Ulva lactuca Linnaeus *(Pl. 2)*

Description: *Ulva lactuca* is one of the most widespread of the five species of *Ulva* in Northwest waters. The thallus of *U. lactuca,* or sea lettuce, is bright green in color. It consists of a single moderate-sized blade (10 to 15 cm) which may be entire or have several splits. Sometimes the margin of the blade is ruffled. A small discoid holdfast anchors the plant to the substrate. *Ulva* plants are always two cells thick. They begin life as small filamentous plants which gradually develop a tubular shape. Although it has walls only one cell thick, the tube collapses so that its sides appear squashed together, forming a broad two-celled sheet in the mature plant. Some species of *Ulva,* such as *U. fenestrata,* have numerous holes or fenestrations in the blade.
Habitat and Distribution: *U. lactuca* grows attached to rocks or other algae in the upper intertidal zone. It may also be found as free-floating plants or fragments of plants in quiet bays, lagoons, or in mud flat areas. *U. lactuca* occurs from Alaska to Chile on the Pacific Coast and is also common in other parts of the world.
Comments: *Ulva* is so widespread that it is familiar to people in many parts of the world and has come to be called sea lettuce. In many countries, it has been used as a condiment in salads, soups, and other dishes. Where *Ulva* occurs in sufficient abundance, coastal farmers have also used it as manure and animal fodder. *Ulva* has an alternation of morphologically similar sexual and asexual plants. Often gamete or spore release coincides with the times of low tides. When *Ulva* becomes fertile, the contents of all the cells at the edge of a blade

will be converted to small, swimming reproductive cells. These cells may form a pale green cloud if gamete or spore release occurs in a still, undisturbed tide pool. When the contents of all the reproductive cells have been released, the margin of the blade will often appear white.

Enteromorpha intestinalis (Linnaeus) Link *(Fig. 11)*

Description: Most *Enteromorpha* plants form hollow tubular plant bodies. In *E. intestinalis,* the tubular shape is quite obvious, especially if the plants are floating in water. These plants typically occur in groups. The entire plant mass may be bright green to almost yellow-green in color. Occasionally, large masses of *E. intestinalis* will become bleached and appear as a white, slightly frothy mat resembling wet tissue paper. Plants typically range from 1 to 20 cm in length and are some 3 to 5 mm in diameter near the broadest part, which is usually at the tip. Often, the ends of the plants will be quite frayed. The walls of the tube are just one cell thick, and the whole tube tapers to a narrow basal or stipe region where it is anchored by a tiny holdfast. *E. intestinalis* may be unbranched or have only a few tubular branches. Other species of *Enteromorpha,* such as *E. prolifera (pl. 2),* may have very narrow tubular thalli with many tubular branches. Careful examination may be necessary to distinguish such *Enteromorpha* plants from filamentous algae with large cells, such as *Urospora.* One interesting and fairly common species of *Enteromorpha, E. linza,* has a tubular basal region, but the more distal part of the plant is quite broad. In fact, it is a large, flattened tube. Unless you look carefully at the basal part, you may mistake *E. linza* for a species of *Ulva*. Altogether, there are some nine species of *Enteromorpha* reported for Northwest waters. They are often very difficult to identify correctly.

Habitat and Distribution: *E. intestinalis* is found attached to rocks in the upper intertidal zone, especially in tide pools or areas with substantial freshwater seepage. Most species of *Enteromorpha* tolerate wide variations in salinity and will grow in water that is only slightly brackish. *E. intestinalis* is known all along the Pacific Coast and is quite common in other parts of the world. Many of the other species of *Enteromorpha* likewise have a cosmopolitan distribution.

Comments: A rather remarkable tolerance to a wide range of salinities, light intensity, temperatures, pollution, and other environmental variables probably accounts for the widespread occurrence of *Enteromorpha*. In many parts of the world, it has been used in a number of recipes, but, since *Enteromorpha* often grows luxuriantly in very polluted waters, one should be very careful when experimenting with such dishes. *Enteromorpha* photosynthesizes and grows quite rapidly. It is not uncommon to see the tubular thalli buoyed by bubbles of photosynthetically produced oxygen trapped in the tubular plant body.

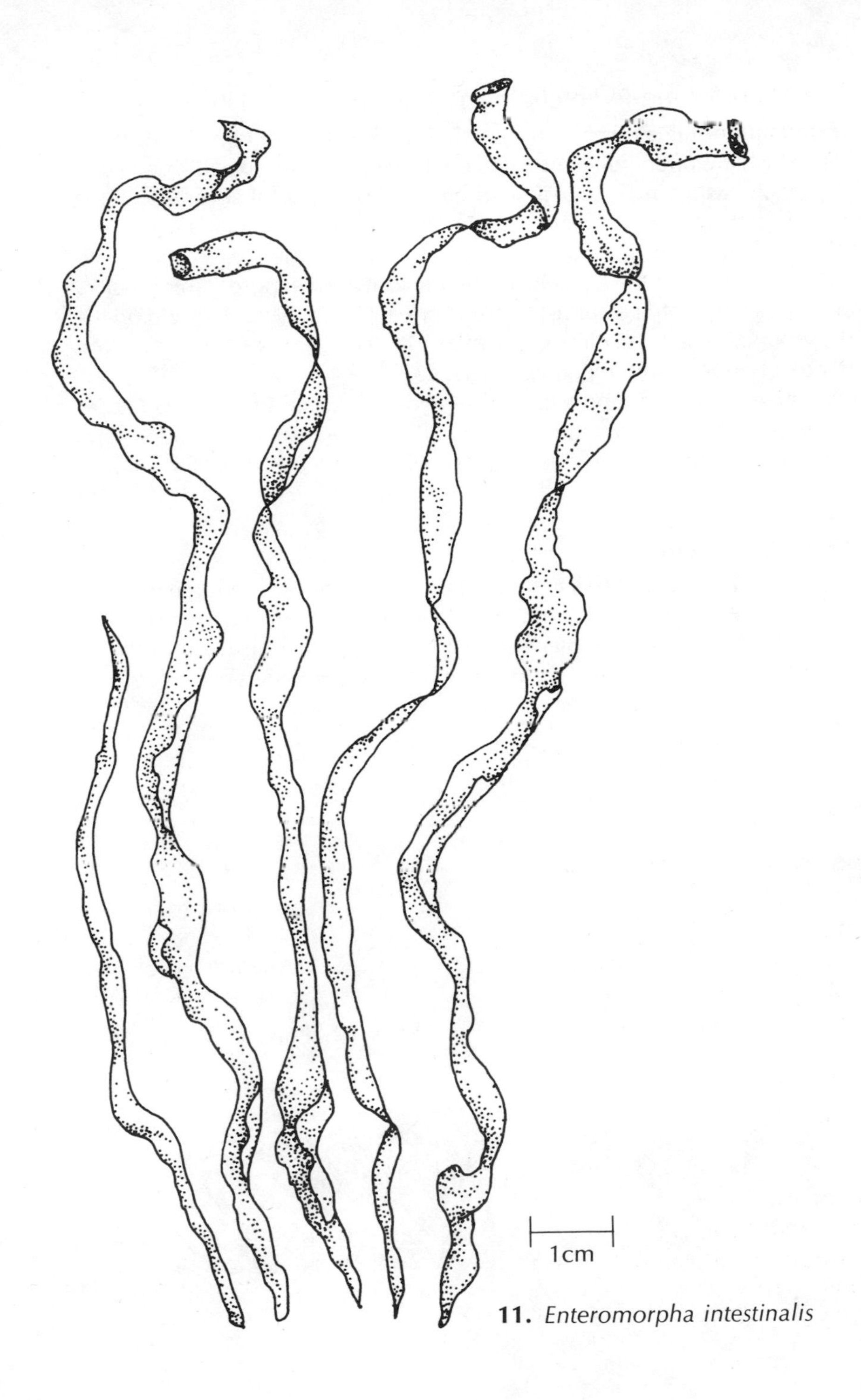

11. *Enteromorpha intestinalis*

Prasiola meridionalis Setchell and Gardner *(Fig. 12)*

Description: This alga consists of very short (0.5 to 1.5 cm high), dark green blades which are often quite curly and contorted. Microscopically, each cell of the blade has a single, star-shaped chloroplast—one of several features which distinguish these tiny blades from other bladelike green algae.
Habitat and Distribution: *Prasiola* is usually found in the spray zone above the highest reaches of the tides. It commonly grows on exposed "bird rocks" or on pilings frequented by birds. Soft, dark green "carpets" of *P. meridionalis* occur from Alaska to central California. Other species of this genus are common elsewhere in temperate waters.
Comments: *Prasiola* is reported to have a very unusual life history. Under certain conditions, it grows into forms that have been described under other names. In mature form, it may be confused with young plants of *Monostroma* or *Ulva*.

Urospora mirabilis Areschoug *(Pl. 3)*

Description: *Urospora* plants are filamentous and dark green in color; they usually occur in groups which resemble dark green streaks or a fringe of thin, slick, green hair on rocks or pieces of wood. Each plant consists of a row of cylindrical to barrel-shaped cells (0.1 mm long by 0.5 mm in diameter in *U.*

12. *Prasiola meridionalis*

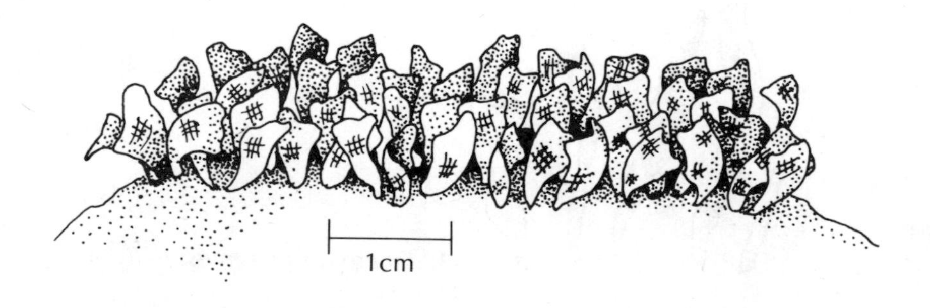

mirabilis, but as much as 1 mm in diameter in some species), joined end to end to form an elongate, unbranched filament (1 to 10 cm long). A number of cells with elongate rhizoidal portions anchor the plant to the substrate. Although there are four species of *Urospora* in Northwest waters, it is very difficult to distinguish one from another.

Habitat and Distribution: Look for *Urospora* on rocks and old logs or on pilings in the upper intertidal zone, especially in areas exposed to moderate wave action. On the Pacific Coast, *U. mirabilis* is found from Alaska to southern California. It is also known in other parts of the world.

Comments: Microscopic examination of recently collected *Urospora* filaments frequently reveals stages in the production of small swimming spores. These have four flagella and a long tapering tail, a characteristic feature of this genus. The spores swim out through a pore in the wall of the parent cell which has differentiated from a vegetative cell into a sporangium. Many aspects of the life history of *Urospora* need further research to clarify all the details.

Cladophora trichotoma (C. Agardh) Kützing *(Pl. 3)*

Description: Of the nine species of *Cladophora* found in Northwest waters, *C. trichotoma* is perhaps the most commonly encountered and most easily recognized. This species forms green hemispherical cushions or tufts some 2 to 5 cm high and 3 to 6 cm in diameter. Close examination reveals that the tufts are composed of much-branched filaments of cells about 0.1 to 0.2 mm in diameter by 0.4 to 1.0 mm long.

Habitat and Distribution: *Cladophora trichotoma* is a conspicuous inhabitant of rocky tide pools and reefs in the upper intertidal zone on more exposed stretches of the coast. It occurs on the Pacific Coast from British Columbia to Mexico and is found in other parts of the world as well.

Comments: *Cladophora trichotoma* is but one representative of a large and widespread genus of green algae. Many of the species form long, tangled tufts, and many have a much finer texture than does the example discussed here.

Acrosiphonia coalita (Ruprecht) P. Hudson *(Pl. 3)*

Description: There are a number of species of this rather ropelike green alga which are only distinguished by microscopic examination. *A. coalita, A. spinescens,* and *A. mertensii* are the most commonly encountered species. The plants are typically 10 to 20 cm long and are composed of highly tangled, branched filaments, which are in turn composed of cylindrical cells approximately 0.1 mm wide by 0.2 to 0.3 mm long. Characteristically, there are also curved, hook-shaped cells or branches which help to quickly distinguish *Acrosiphonia* from *Cladophora,* another common filamentous green alga. You will need the aid of a hand lens or microscope to see these "hooks." Often, *Acrosiphonia* plants form a fairly extensive fringe or mat on intertidal rocks. In sheltered areas, the plants may not appear bright green unless they are fluffed up, since they typically get covered by mats of golden brown diatoms.

Habitat and Distribution: *Acrosiphonia* occurs on rocky shores in the lower intertidal zone of the sheltered inland waters of the Northwest. Somewhat sheltered sites on more exposed parts of the coast may also harbor some *Acrosiphonia*.

Comments: Like many other marine algae, the plant that is readily seen and collected in the field tells only half the story. The large ropy plants (formerly called *Spongomorpha)* you will find in nature are the sexually reproducing phase of this seaweed. Sexual reproduction, which occurs by fusion of tiny swimming gametes, ultimately results in the formation of a tiny (less than 1 mm) green cell which can be found in the blades of such red algae as *Porphyra* and *Schizymenia,* or in the tissues of the rock-encrusting red alga, *Petrocelis*. Before their relationship to *Acrosiphonia* was discovered, these tiny green cells were named as separate entities *(Codiolum* or *Chlorochytrium).* Even without a microscope, the practiced eye can sometimes discern these tiny endophytes which appear as tiny green dots within the tissues of the host algae. While they are only found in association with other algae in nature, they can be cultured free of their hosts in the laboratory. These *Chlorochytrium* or *Codiolum* phases of *Acrosiphonia* produce numerous swimming spores which germinate into the ropelike *Acrosiphonia* plants.

Derbesia marina (Lyngbye) Solier *(Pl. 4)*

Description: Spore-producing plants of *Derbesia marina* consist of fine, branched, tubular cells without cross walls (0.5 mm in diameter and up to 1 to 4 cm long). They are rarely conspicuous in nature and are usually only found after careful searching. More conspicuous than the asexual *Derbesia* stage of this plant is the gamete-producing, sexual stage, once known as *Halicystis ovalis*. This stage consists of globose vesicles (each one a single cell up to 1 cm in diameter) which resemble dark green pearls. These are anchored to the substrate by a thin rhizoid which penetrates the substrate.

Habitat and Distribution: Both the asexual *Derbesia* stage and the sexual "*Halicystis*" stage of this remarkable plant can be found from Alaska to Mexico in the lower intertidal or in the subtidal zone to a depth of at least 10 m. The *Derbesia* stage has been found on red coralline algae, sponges, and rocks, while the "*Halicystis*" stage is only encountered on encrusting coralline red algae.

Comments: The chief reason for mentioning this species—which may be difficult to find—is to include an example of the remarkably large size reached by a single plant cell. The large, globose "*Halicystis*" plants are especially spectacular and quite sturdy. I once found great numbers of them washed up intact on a sandy beach adjacent to a rocky reef. *Derbesia* is also an excellent example of an alga with two very different stages in its life history—the filamentous *Derbesia* stage (spore-producing) and the globose *"Halicystis"* stage (gamete-producing). Long before algologists were able to study the life history of these two different looking plants, they were named as separate species. Only through careful laboratory study was it discovered that these two species were really only different phases of the one species.

Codium fragile (Suringar) Hariot *(Pl. 4)*

Description: *Codium fragile* is a dark green alga which ranges from 10 to 40 cm in height and consists of repeatedly branching cylindrical segments about 0.5 to 1.0 cm in diameter. These segments look like dark green fingers. The holdfast is a broad, spongelike cushion of tissue. The plant body actually consists of a number of interwoven, filamentous cells with incomplete crosswalls forming the inner part of the branches. The surface layer of the branches is formed by elongate, club-shaped vesicles upon which the gametangia are produced. Often, the surface of the plant may be covered by a "halo" of other small epiphytic algae.

Habitat and Distribution: *Codium fragile* inhabits the middle and lower intertidal zone as well as the subtidal regions of rocky shores. It can also be found in large tide pools which are permanently filled with water. It occurs from Alaska to Mexico, and in many other parts of the world. A second species of *Codium, C. setchellii,* also occurs in regional waters, it resembles a dark green sponge and forms a mat 1 cm thick and 2 to 10 cm in diameter which has no upright, branching part.

Comments: While *C. fragile* is apparently a long-time resident of this coast, it invaded New England coastal waters about 1957 and has spread rapidly since then. In New England, new plants of *C. fragile* often start in mussel, scallop, or oyster beds, where the shells of these animals provide a firm attachment surface for the small plants which then grow rapidly. However, once the plants reach a large size, waves surging through the shellfish beds tend to move away the alga and its host shellfish from their normal habitat. This often spells doom for the shellfish; thus, the *Codium* invasion has proved to be a serious problem for the New England shellfish industry. A number of studies are underway at university and government laboratories in an effort to understand and control the wanderings of this truly "weedy" seaweed.

Bryopsis corticulans Setchell *(Fig. 13)*

Description: When matted together in a clump, the featherlike plants of *Bryopsis* appear dark green; if submerged, they appear bright green and their featherlike appearance is quite striking. Usually only 4 to 5 cm high, these plants may grow as long as 15 cm. The plant body consists of a central tubular axis 1 mm in diameter, with slightly smaller branches produced mostly in one

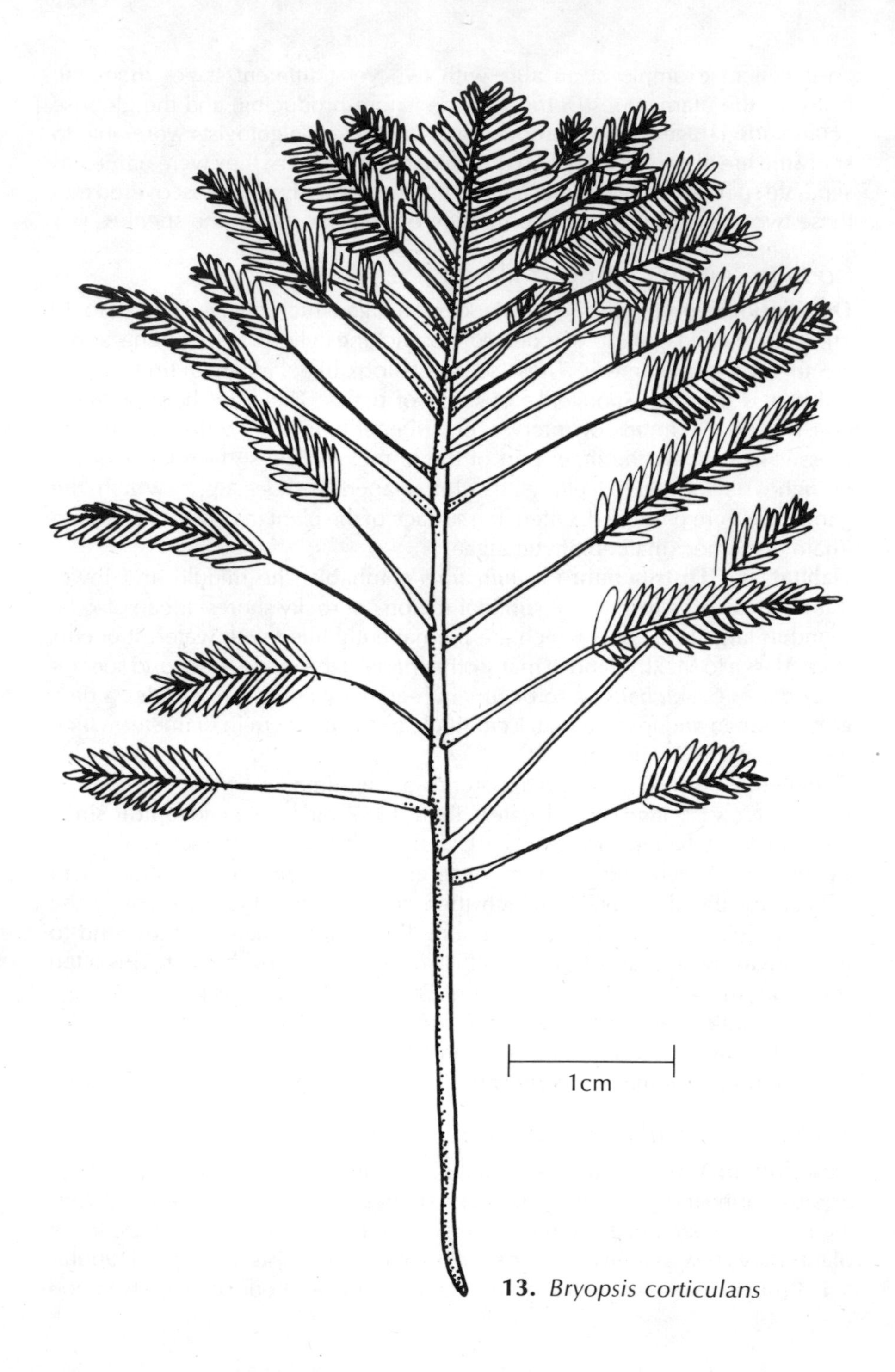

13. *Bryopsis corticulans*

plane; the length and size of the small branches decrease near the apex of the plant.
Habitat and Distribution: *Bryopsis* is found on rocks, shells, and wood in the lower intertidal and subtidal zones from British Columbia to Mexico.
Comments: *Bryopsis* is another example of a marine alga with very large cells, for most *Bryopsis* plants consist of only one or a few large, highly branched cells which are easily seen with the naked eye.

Brown Algae (Phaeophyta)

This group of algae is the most exclusively marine of all the major groups. Only three genera of brown algae occur in fresh water. Brown algae contain the green pigment chlorophyll. However, the chlorophyll is well masked by an abundance of other pigments which are gold or brown in color. Hence, these algae may range in color from a light olive green to a dark, almost black, chocolate brown. Photosynthetically produced food reserves are stored in the form of a soluble polymer called laminarin. The cell walls contain cellulose and other polymers, such as algin. While in most groups of algae there are some species which are never more complex than a single cell, the simplest brown algae are multicellular filamentous plants. However, all brown algae have some single-celled spores or gametes and many of these are able to swim around under their own power for a short time.

The largest and most conspicuous seaweeds are found among the kelps which are brown algae. Many of these kelps are sources of useful products, serving as the raw materials for a sizable industry. There are about fifty genera and perhaps 140 species of brown seaweeds in Pacific Northwest waters. A number of these, especially the more conspicuous ones, are described and illustrated in the following pages.

Ectocarpus siliculosus (Dillwyn) Lyngbye *(Fig. 14)*

Description: *Ectocarpus* is a branched filamentous alga which forms soft, light brown tufts or plumes on rock, wood, or other algae, especially the larger brown algae. The tufts are typically 1 to 2 cm long but may be longer. A microscope is needed for critical identification of *Ectocarpus* species, there being about a dozen species in this region. In addition, there is a closely related genus, *Pilayella,* which resembles *Ectocarpus;* it is distinguished by the location of its sporangia or gametangia. Individual cells of *Ectocarpus* are only 0.02 to 0.05 mm in diameter and 1 to 2 times as long as they are wide. *Ectocarpus* has elongate multi-chambered reproductive organs which may be gametangia or sporangia. It also has ovoid, single-chambered sporangia. These structures are borne on short side branches.
Habitat and Distribution: *Ectocarpus* grows on rocks or wood, as well as on other larger, brown algae in the lower intertidal zone. On the Pacific Coast, *E.*

14. *Ectocarpus siliculosus* -Habit

siliculosus occurs from Alaska to central California; it is widespread on the Atlantic Coast, as well. Other species of *Ectocarpus* and closely related genera are prevalent in most waters of the world and are often more conspicuous than in the Northwest.

Comments: *Ectocarpus* is included here as an example of a common and widespread group of small brown algae of fairly simple construction. There are many species of brown algae with this general type of organization; all require careful microscopic examination and measurement for identification.

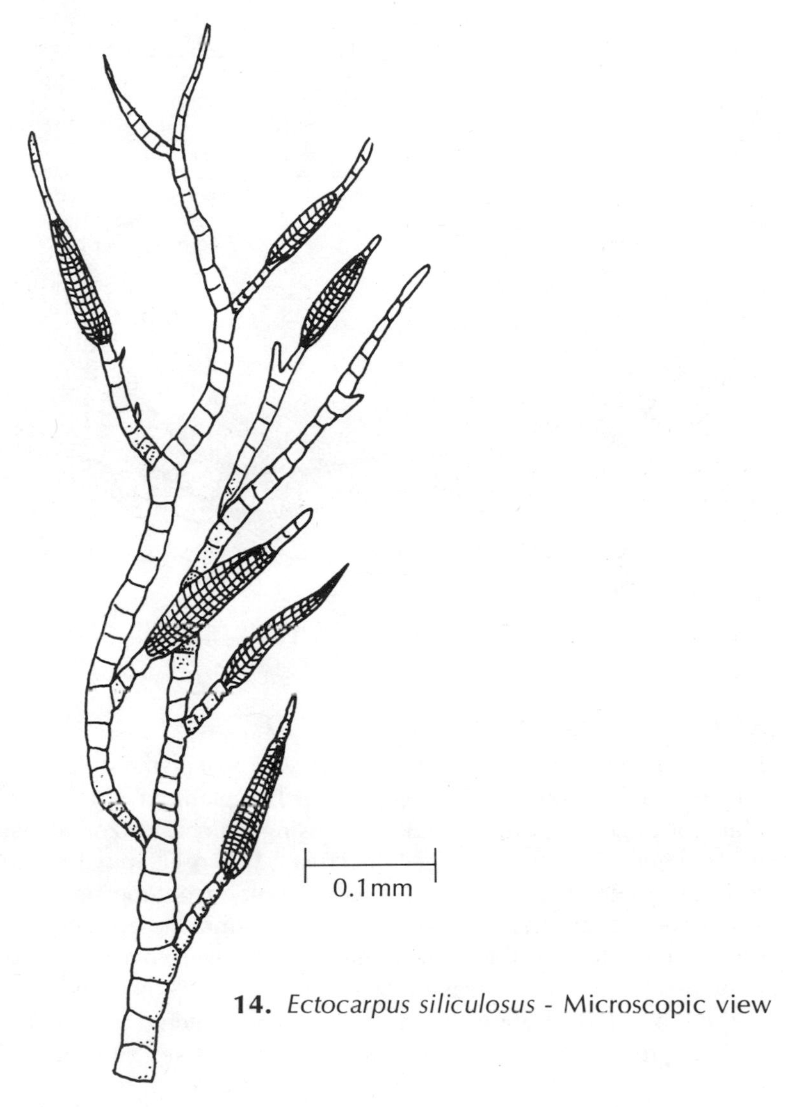

14. *Ectocarpus siliculosus* - Microscopic view

In searching for *Ectocarpus* and related genera in the protected inland waters of the Northwest, you will undoubtedly pick up species of filamentous diatoms which are usually more abundant than *Ectocarpus* and its relatives. Often these filamentous or colonial diatoms are darker than the filamentous brown algae, and the tufts disintegrate more easily when rubbed between the fingers than do the filamentous brown algae. A good hand lens or low-powered microscope will aid in distinguishing the filamentous brown algae from these diatoms.

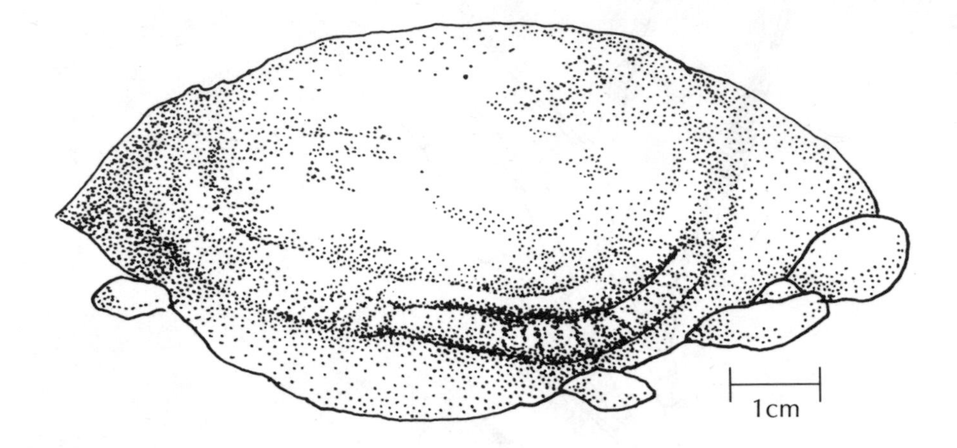

15. *Ralfsia pacifica* -Habit

Ralfsia pacifica Hollenberg *(Fig. 15)*

Description: *Ralfsia* forms a thin, dark brown crust on the surface of rocks and has received the common name tar spot. The plant consists of short upright filaments, tightly packed together and rising from a horizontal layer of cells located next to the rock surface. The crust is 0.5 to 1.0 mm thick and may be from 2 to 10 cm in diameter. It is usually circular in outline in small specimens and lobed or wavy in large specimens. To the unaided eye, the surface of the crust appears to have fine radial ridges emanating from a central point. There may also be circular or wavy ridges which parallel the edge of the crust.

Habitat and Distribution: *Ralfsia* is often encountered high in the intertidal zone on rocks and other solid surfaces. This species has been found from Alaska to Baja California.

Comments: There are other species of algae which resemble *Ralfsia*; some brown algae, such as the tubular *Scytosiphon (pl. 5),* have crustose phases in their life histories which are easily mistaken for *Ralfsia*. There are also a number of species of red algae, such as *Petrocelis,* which form dark crusts on intertidal rocks. Such encrusting algae appear to survive both high temperatures and exposure better than some of the tubular or bladelike forms which inhabit similar levels in the intertidal region and often succumb to hot summer temperatures.

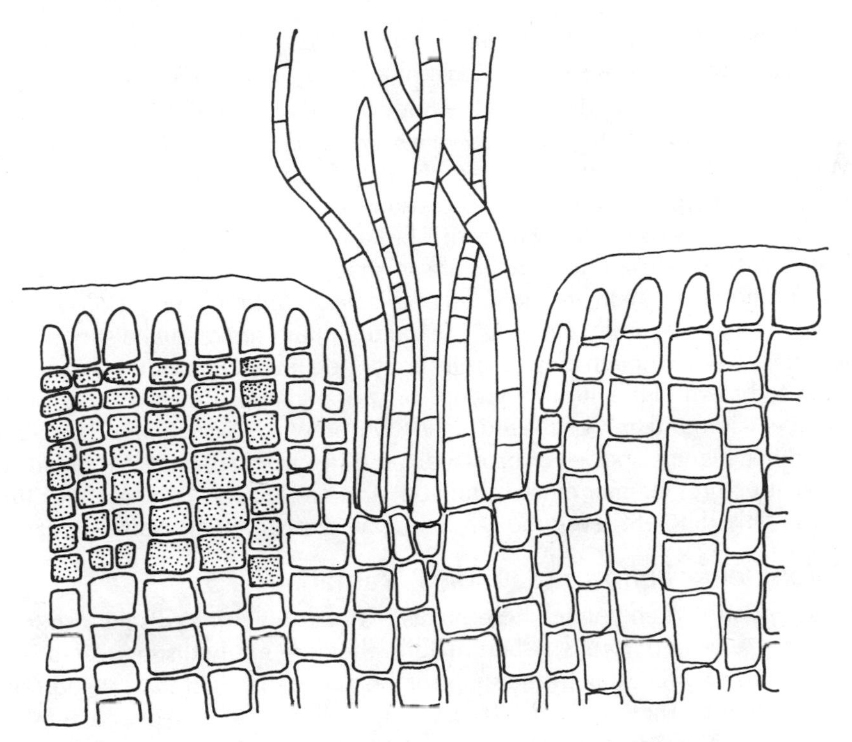

15. *Ralfsia pacifica* - Vertical microscopic section

15. *Ralfsia pacifica* - Surface view, microscopic

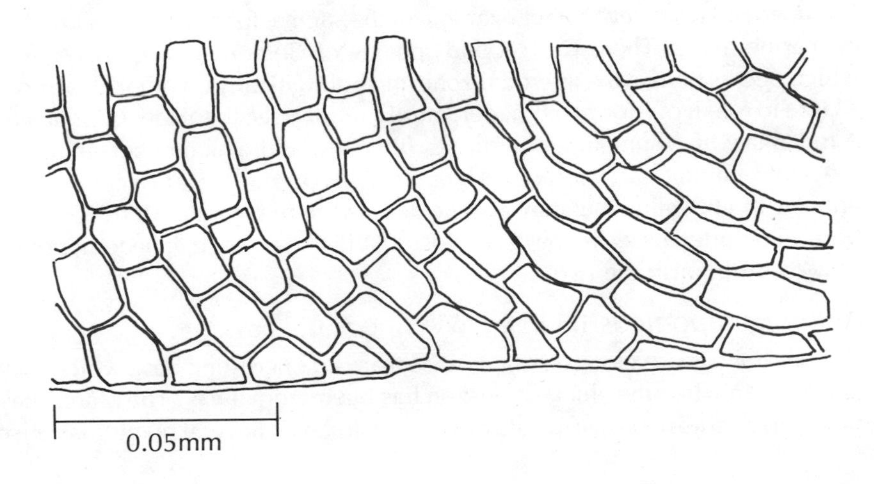

Leathesia difformis (Linnaeus) Areschoug *(Pl. 5)*

Description: This plant forms small spherical sacs about 1 cm in diameter which grow to large and flattened, spongy, lobed or convoluted sacs between 5 and 6 cm in diameter. These seaweeds are light brown to yellowish brown. When squashed between the fingers, they will break into masses of filaments.
Habitat and Distribution: *Leathesia* is a fairly common intertidal alga which grows on rocks or other algae from Alaska to Mexico.
Comments: There are a number of sac-shaped brown algae in Northwest waters; not all of them are included in this book. When young, *Colpomenia bullosus* may resemble *Leathesia,* but mature plants are readily distinguished in the field even though they inhabit similar tide levels. *Soranthera* is another saccate brown alga with tiny groups of sporangia appearing as dots on its surface; it is found only on the red algae *Rhodomela* or *Odonthalia*. These and still other saccate species of brown algae differ from *Leathesia* in details of body structure, manner of reproduction, and life history. They are thus fundamentally different, even though they may appear superficially similar.

Colpomenia bullosus (Saunders) Yamada *(Pl. 5)*

Description: When young, these plants may be difficult to distinguish from young *Leathesia difformis;* when mature, they are easily distinguished by a very smooth surface and by the shape of the mature, upright, saccate portions of the plants. They are 1 to 10 cm high, slightly flattened and twisted. Yellowish brown in color, they arise from a contorted, squashed-looking basal portion. When pulled or broken, they tend to tear rather than disintegrate, as does *Leathesia*.
Habitat and Distribution: *C. bullosus* forms little clusters of plants on rocks in the mid-intertidal region from Alaska to Mexico. It is usually not as abundant as *Leathesia*.
Comments: This is yet another example of the saclike form found in a number of marine algae. There is a second species of *Colpomenia, C. peregrina,* which occurs both in the intertidal zone and epiphytically on other algae from Alaska to Mexico. It forms a broader, more rounded sac than does *C. bullosus*. A red alga which sometimes mimics *C. bullosus,* and which occurs in nearly the same habitat, is *Halosaccion glandiforme (fig. 30)*. *Halosaccion* often appears bright yellow but may also be deep wine red in color. It never arises from a contorted base, as does *C. bullosus*. With experience, it becomes quite easy to distinguish the two.

Analipus japonicus (Harvey) Wynne *(Fig. 16)*

Description: These elongate plants arise from an encrusting base which may persist even when the elongate portion has been worn away. The plants may range from reddish brown to pale olive green in color.Several plants may arise

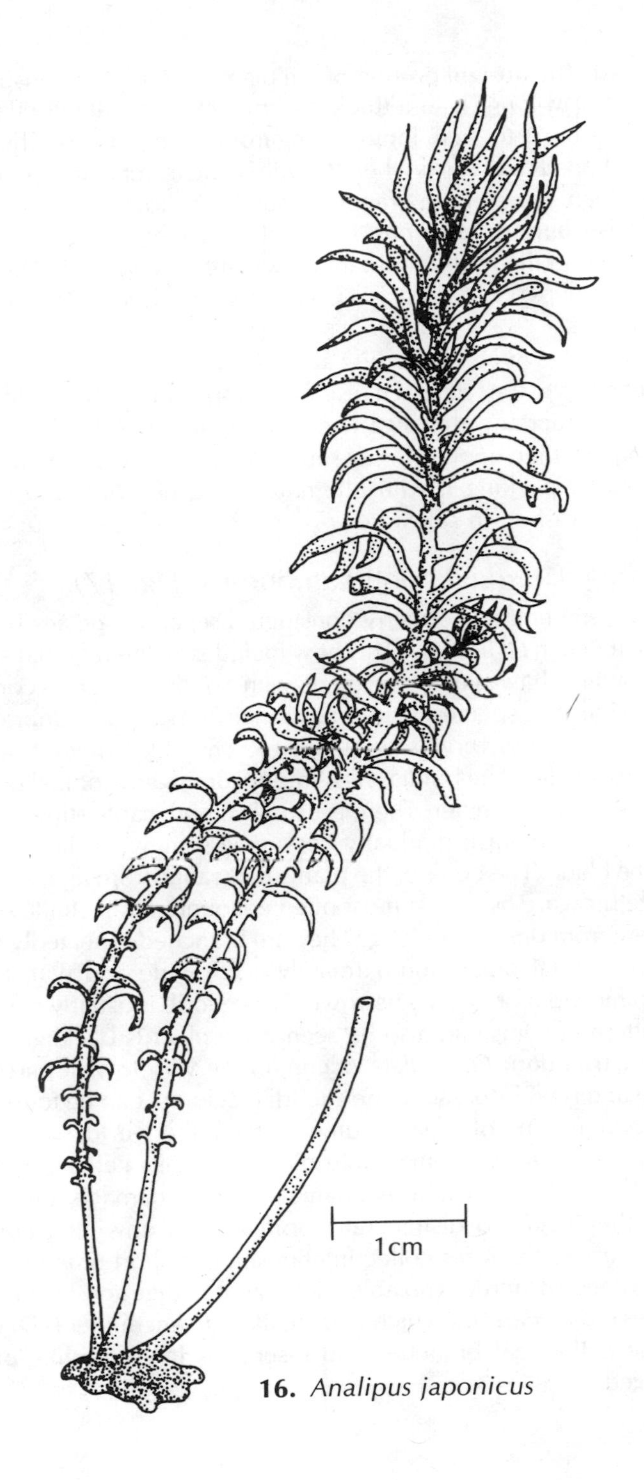

16. *Analipus japonicus*

from one holdfast. The upright portion of the plant consists of a conspicuous main axis 5 to 25 cm long, with a thick covering of short, slightly flattened lateral branches (from 1 to 3 cm long) arising from the main axis. The plants are often found draped over rocks when the tide is out. In very large plants, the main axis and even some of the branches may be hollow.

Habitat and Distribution: *Analipus* is found on rocks along moderately exposed shorelines in the mid-intertidal to lower intertidal zones. The plants usually occur in groups. On the Pacific Coast, *Analipus* is found from Alaska to central California.

Comments: This alga was formerly called *Heterochordia abietina;* the specific epithet refers to the fact that it resembles a sprig of fir *(Abies)*. In Japan, where it is quite abundant, this alga is called *matsumo*. A food item, it is preserved in salt and then cooked in soy sauce before eating. Salted *matsumo* is also prepared by packing it with alternate layers of edible mushrooms which have been washed in salt water.

Desmarestia ligulata (Lightfoot) Lamouroux *(Fig. 17)*

Description: Several entities formerly considered separate species have recently been included in *D. ligulata*. This now includes all the regional species of *Desmarestia* which have a flattened blade with a midrib. There is considerable variation within this species: plants may range from a single, unbranched blade (up to 60 cm tall) to very broadbladed (40 cm wide), much-branched plants with many blades. The holdfast is a smooth or ridged, conical or lobed structure up to 4 cm in diameter. There is usually a noticeable stipe which is cylindrical near the holdfast, gradually becoming somewhat flattened and immersed in the blade. The color of the plants may range from light brown to rich, chocolate brown. The plants most often encountered in shallow nearshore waters are from 0.5 to 1 m long. They are branched repeatedly with a central, flattened, axial blade approximately 2 cm wide and with branch blades of a similar width or slightly narrower. The midrib is usually conspicuous, and a pattern of veins can also be seen if the plant is backlighted.

Habitat and Distribution: *D. ligulata* occurs in the very lowest part of the intertidal zone and is widespread in the subtidal region. It can be found from Alaska to Mexico and in cold seas in other parts of the world.

Comments: *Desmarestia* is a remarkable plant in that its cell sap contains sulfuric acid. If the plants sustain mechanical or heat damage, the acid is released, and they begin to disintegrate rapidly. If an unwitting collector places some *Desmarestia* in his collecting bucket, in a short time his whole collection becomes an unrecognizable mess, as it is digested by the acid. Another species of *Desmarestia* which you are likely to encounter is *D. virdis,* which has fine cylindrical branches and resembles large hairlike plumes when submerged.

17. *Desmarestia ligulata*

Scytosiphon lomentaria (Lyngbye) J. Agardh *(Pl. 5)*

Description: *Scytosiphon* is a tubular brown alga which may measure up to 50 cm in length and 0.5 to 1.0 cm in diameter. The plants, which often grow in groups, are olive brown to dark brown. Sometimes the cylindrical plant body is slightly constricted at regular intervals, so that it resembles a series of small link sausages.
Habitat and Distribution: *Scytosiphon* can be found in the mid- to lower intertidal zone and on floating docks or log booms. It ranges from Alaska to Mexico.
Comments: When lying in calm sunlit water, the hollow plant body often becomes filled with gas bubbles as a result of rapid photosynthesis.

Petalonia debilis (C. Agardh) Derbes and Solier *(Pl. 6)*

Description: *Petalonia* is a small blade 5 to 30 cm long by 2 to 10 cm wide. It has a discoid holdfast and a narrow stipe which broadens into a blade. The blade, pointed at the apex, has a somewhat irregular outline with wavy edges. It varies in color from dark brown to pale, almost greenish brown. There are a number of brown algae which can only be distinguished from *Petalonia* by microscopic examination. If examined in cross section, *Petalonia* has a surface layer of small, pigmented cells and a central core of larger, slightly pigmented cells.
Habitat and Distribution: From Alaska to Mexico, *Petalonia* is found on rocks in the intertidal zone or as an epiphyte on eelgrass, *Zostera,* or surfgrass, *Phyllospadix.*
Comments: *Petalonia* is but one of many epiphytes you may find growing on the surface of the green strap-shaped leaves of the flowering sea grasses, *Zostera* and *Phyllospadix.* Other algal species, including the green alga, *Monostroma zostericola,* and the red alga, *Smithora naiadum,* can be found growing as epiphytes on these aquatic flowering plants.

Laminaria saccharina (Linnaeus) Lamouroux *(Pl. 6)*

Description: One of the most commonly encountered species of *Laminaria* in the Pacific Northwest, *L. saccharina* has a long brown blade of variable length (up to 2 m or more) and width (12 to 20 cm). The blade is often split or torn longitudinally along lines which reflect the location of mucilage ducts. The surface of the blade may be smooth or have corrugations which run in two rows paralleling the edge of the blade, like ruffles. The base of the blade may narrow to a broad wedge shape or be broadly rounded where it connects with the slightly flattened, flexible stipe. The stipe may be from 10 to 40 cm long and arises from a holdfast with numerous finely branched haptera. The plants are usually a rich brown color. Older plants, especially those in relatively

calm waters, may be heavily encrusted with growths of small algae and animals.
Habitat and Distribution: *L. saccharina* can be found growing on shells, wood, and rocks in moderately to very sheltered waters between lower intertidal and upper subtidal regions. Where heavy surf occurs, the plants are restricted to the subtidal zone. On the Pacific Coast, this species of kelp ranges from the Aleutian Islands to Coos Bay, Oregon; an isolated population has also been reported at Santa Catalina Island, California. *L. saccharina* is also widespread on North Atlantic shores.
Comments: This is the first of several kelps described and illustrated in the following pages. The word *kelp* is derived from a Middle English word referring to the ash made by burning brown algae of the orders Laminariales and Fucales. This ash provided a potash useful in making soap and in other industrial processes. Later, kelp was found to be a good source of iodine, and for many years before the discovery of iodine-rich mineral deposits in Chile, it was the principal source of iodine. The vast majority of kelps are large and easily visible seaweeds in the cold and cold-temperate seas of the world. The Pacific Coast has perhaps the greatest diversity of species to be found anywhere—with nineteen genera and forty species reported from Alaska to Baja California. About 80 percent of these species can be found in Pacific Northwest waters. Besides their usefulness as a source of chemicals, many kelps provide food and shelter for marine animals, and some are eaten by man as flavoring in soups, stews, and similar dishes.

Laminaria groenlandica Rosenvinge *(Pl. 6)*

Description: The blade of this species of *Laminaria* is deep rich brown to almost black in color, measuring from less than 1 m to as much as 2 m long by 10 to 50 cm wide. The blade may be entire or have longitudinal splits; it may also have corrugations. The base of the blade is broadly tapered or rounded. It arises from a slightly flattened, flexible stipe, which may measure from less than 10 to up to 60 cm long. The holdfast is easily noticeable with many finely branched haptera.
Habitat and Distribution: *L. groenlandica* grows on rocks in the lower intertidal or upper subtidal zone, especially in areas with substantial water motion. In sheltered waters, it is restricted to the subtidal region. This species occurs from Alaska to southern Oregon.
Comments: In addition to the two species of *Laminaria* mentioned here, there are eight other species in Pacific Northwest waters. Some are quite variable in form, and many require careful examination for critical identification.

Agarum cribrosum Bory *(Pl. 7)*

Description: *Agarum* plants consist of a large (50 to 90 cm long), broad (20 to

50 cm wide) blade with a smooth, somewhat ruffled margin. A midrib 1 to 3 cm wide arises near the holdfast and runs up the center of the blade. The blade is dark brown and has numerous perforations; it tears easily and may quickly turn green when removed from the water. The stipe is cylindrical, 3 to 30 cm long, and only 5 to 8 mm in diameter. The holdfast is small with numerous fine branches.

Habitat and Distribution: *A. cribrosum* is found on rocks from the upper subtidal zone to depths of 10 m. It often forms extensive underwater "meadows" and is frequently seen by divers. *A. cribrosum* has been found from the Bering Sea to northern Washington.

Comments: While kelps are usually the favorite food of sea urchins, a study conducted in Washington indicates that sea urchins find *Agarum* distasteful; consequently, *Agarum* does not suffer as much grazing pressure as other kelps and can survive and form extensive populations, even though it has a slower growth rate than most other kelps. A second species of *Agarum, A. fimbriatum,* has few perforations in its blade and a more corrugated blade with a toothed margin. *A. fimbriatum* occurs in the lower intertidal and subtidal zones in rocky areas from southern British Columbia to southern California.

Hedophyllum sessile (C. Agardh) Setchell *(Pl. 7)*

Description: While most species of kelps have conspicuous stipes, *Hedophyllum* is notable because it lacks a stipe when mature. Thus, *Hedophyllum* appears as a clump of smooth or corrugated blades from 30 cm to more than 1 m long, which appear to spring directly from the rocks. The light to very dark brown blades are often deeply split into 5 to 10 cm wide segments, especially where water motion is extreme. By carefully pushing back the blades, one can see the numerous haptera of the holdfast arising directly from the base of the blade. These are usually hidden by the canopy of blade segments.

Habitat and Distribution: *Hedophyllum* is quite common on rocks in the mid- and lower intertidal regions from Alaska to northern California.

Comments: *Hedophyllum* is a good example of an alga which exhibits different growth forms depending on the amount of water motion it experiences. In the more sheltered inland waters of the Northwest, these plants typically have broad, highly corrugated blades with few longitudinal splits. On the wave-swept outer coast, the blade segments are usually narrow, smooth and strap-shaped. The corrugated form is believed to offer greater resistance to water motion and might be torn apart more readily than the smooth form. A similar pattern is also observed in the amount of corrugation present in the blades of other species of kelps.

Nereocystis luetkeana (Mertens) Postels and Ruprecht *(Pl. 7)*

Description: *Nereocystis* is one of the largest seaweeds known. Up to 40 m long, it is quite common and noticeable in Pacific coastal waters. It grows in groves offshore, its most visible part being an inflated float (10 to 15 cm in diameter) at the top end of the stipe. The float supports a profusion of long, strap-shaped blades (3 to 4 m by 10 to 15 cm) which trail in the water. The blades arise from two principal regions on top of the float. The float is attached to a long, hollow stipe which is usually between 5 and 15 m long, and the whole plant is anchored to the substrate by a massive holdfast up to 40 cm broad, with many overlapping haptera. During summer and early fall, the long strap-shaped light brown or golden colored blades often have chocolate brown colored patches which are groups of sporangia (each group of sporangia is called a *sorus*); millions of swimming spores are produced by each *Nereocystis* plant. Some of these manage to settle on the rocky bottom where they germinate and grow into microscopic, filamentous plants which reproduce sexually. Following fertilization of an egg by a sperm (which probably occurs in winter), a new young spore-producing phase develops and begins to grow rapidly in the spring. In only one growing season, these young plants become the massive *Nereocystis* plants which are so common in local waters during spring and summer. The majority of these large annual plants are usually swept away by fall and winter storms; they drift out to sea or are cast ashore where they die.

Habitat and Distribution: *Nereocystis* forms conspicuous groves offshore in the upper subtidal region and to depths of 10 m or greater in areas with rocky bottoms. Occasionally, small plants may begin to grow in the lower intertidal zone; these intertidal plants usually die off during the daytime low tides of spring and summer. This seaweed is found from Alaska to California.

Comments: Besides providing a habitat for many marine fishes and invertebrates, *Nereocystis* is eaten by some animals, such as sea urchins. Excellent pickles can be made from sections of the cylindrical stipe (see recipe in Chapter VIII). Northwest Indians used long pieces of dried, oiled stipe as fishing line. Although there have been attempts to harvest *Nereocystis* commercially, they have not been economically successful.

Postelsia palmaeformis Ruprecht *(Pl. 8)*

Description: *Postelsia* is the "sea palm" of the Pacific Coast. It does indeed resemble a small, olive brown palm tree some 15 to 50 cm tall. *Postelsia* plants have sturdy holdfasts composed of numerous fingerlike projections which arise at the base of the thick (2 to 4 cm) but hollow upright stipe. Near the top are a number of longitudinally grooved, strap-shaped blades 1 to 3 cm wide and up to 25 cm long. There may be as many as one hundred blades on a single plant. The upright posture of its stipe with its straplike blades hanging down, gives *Postelsia* a palmlike appearance.

Habitat and Distribution: *P. palmaeformis* is found from central California to

Hope Island, British Columbia. Within this range, it occurs only on the most exposed portions of the outer coast, where it forms conspicuous groves on wave-battered reefs in the mid-intertidal zone. It is often found in association with mussels *(Mytilus californianus)* and barnacles *(Balanus cariosus)*. Although you will never find it actually growing in protected or sheltered waters, occasionally you may find it washed ashore following storms.

Comments: *Postelsia* is an annual plant which is most discernible at its maximum size in the summer and fall. Sporangia are located in the grooves of the blades; these release microscopic swimming spores. It is unlikely that the spores swim very far; they may just drip down along the grooved blades onto the rocks below. Such a mechanism would assure persistence of *Postelsia* patches from year to year in the same place on the reef. These spores produce microscopic, branched filamentous plants which overwinter and reproduce sexually, giving rise to the large *Postelsia* plants which you can easily observe in nature. While *Postelsia* spores can successfully colonize bare rock surfaces or barnacle shells, they do not seem to be able to successfully settle in areas where mussels are growing. Sometimes when large growths of *Postelsia* cover a patch of barnacles, the barnacles may weaken and die, and the whole assemblage of dead barnacles and *Postelsia* may be wrenched from the rock by strong waves. The bare rock surfaces cleared by this mechanism can then be the site for the establishment of a longer lasting grove of *Postelsia*.

Macrocystis integrifolia Bory *(Pl. 8)*

Description: This species of kelp forms extensive forests and is one of the largest and most complex algae. Floating at or near the sea surface are a large number of golden to rich-brown leaflike blades 25 to 35 cm long by 5 cm wide. The blade is wrinkled or grooved in an irregular pattern; its edge is lined by toothlike projections. The tips of the blades are pointed, but the bases are broadly rounded and fastened directly to round-to-oval floats which arise from the stipe. Hundreds of such blade-bearing floats may be borne on a single repeatedly branched stipe. *Macrocystis* stipes are about 1 cm in diameter and up to 30 m in length. They arise from a creeping, somewhat flattened, profusely branched rhizomelike base which has many branched haptera. Clusters of smooth-surfaced, toothed blades near the base of the plant are sporangia-bearing blades (sporophylls). The terminal blade located at the tip of the stipe produces new blades along its innermost edge, and it is a good place to see a series of maturing blades and floats.

Habitat and Distribution: *Macrocystis integrifolia* occurs in the very lowest portion of the intertidal and in subtidal waters 7 to 10 m deep. It favors areas exposed to the open sea but somewhat sheltered from the full force of heavy wave action. It does not seem to grow in areas with salinity lower than that of the open coastal waters, so it is not found very far into the Strait of Juan de

Fuca. Where favorable conditions do exist, it is present from Alaska to California. This same species is reported from the cold waters on the west coast of South America; other species of *Macrocystis* are found around islands and land masses in the cold-temperate seas between 30° and 60° south latitude.

Comments: In the waters adjacent to southern California, vast forests of *Macrocystis* are harvested by an underwater mowing machine mounted on a self-propelled barge. This machine only cuts part of the plant within 1 m of the sea surface, leaving most of it to regrow and provide a sustained yield. *Macrocystis* is an excellent source of alginate, a product used to control the texture and consistency of salad dressings and ice cream. Alginate is also used in pharmaceuticals and in the processing of items such as beer, textiles, paint, and paper products.

Pterygophora californica Ruprecht *(Pl. 9)*

Description: This species has a very long, stout, almost "woody" stipe. It is about 3 to 4 cm in diameter, slightly oval in cross section and up to 2 m long. The holdfast is stout with numerous branched haptera. The top of the stipe is flattened and bears two rows of lateral blades and a single terminal blade. The blades are typically 6 to 10 cm wide and 30 to 60 cm long. Their color is dark chocolate brown.

Habitat and Distribution: *Pterygophora* occurs on rocks in the subtidal zone from the upper region to depths of 10 m. In fairly dense stands, it is found from British Columbia to Mexico.

Comments: *Pterygophora* is an excellent example of a long-lived perennial seaweed. The blades usually degenerate or wear away in the winter, and new ones are produced each spring.

Alaria marginata Postels and Ruprecht *(Pl. 9)*

Description: *Alaria* is most easily recognized by its long olive green to rich brown terminal blade (2 to 3 m long) which arises from a short stipe. The stipe is cylindrical near the holdfast but flattens out in the region of the terminal blade to become a conspicuous midrib (5 to 20 mm wide). Frequently, the tip of the blade is quite tattered, and the midrib may actually extend beyond the limits of the flat blade. At the base of the main blade, mature plants will have two opposite rows of chocolate brown sporangia-bearing blades (sporophylls) which are 10 to 20 cm long by 2 to 3 cm wide and somewhat oval in outline. The holdfast consists of many slender branching haptera.

Habitat and Distribution: *A. marginata* is quite common in the lower intertidal zone on moderate to very exposed rocky shores. It is found from the Gulf of Alaska to central California.

Comments: Of the four other species of *Alaria* in Pacific Northwest waters, only *A. nana* is very common. It is much smaller than *A. marginata* and occurs only in the mid-intertidal region of the outer coast where the plants are exposed to extreme surf conditions. Altogether, there are fourteen species of *Alaria* found in the Arctic, North Atlantic, and North Pacific oceans.

Lessoniopsis littoralis (Farlow and Setchell) Reinke *(Pl. 9)*

Description: These plants are typically dark brown in color and up to 2 m in length. Their most noticeable feature is a very stout, very "woody" holdfast which may be as much as 20 cm thick and up to 40 cm long. These holdfasts resemble small tree trunks; older ones are deeply furrowed. The stipe is repeatedly branched, and the tip of each branch bears elongate, flattened blades 7 to 12 mm wide and up to 1 m long. A 2- to 3- mm wide midrib runs down the center of each blade. Old plants may have as many as 500 blades.
Habitat and Distribution: You will find *Lessoniopsis* only in the lowest part of the rocky intertidal zone in areas exposed to the extreme surf of the open coast. Observing in such sites can be quite dangerous, and one must pay constant attention to incoming waves. This species is found from Alaska to central California.
Comments: *Lessoniopsis* is a long-lived perennial, remarkable for its stout construction. It is an excellent example of a seaweed restricted in distribution because of environmental factors, such as tidal height and degree of exposure to wave action.

Egregia menziesii (Turner) Areschoug *(Fig. 18)*

Description: Arising from a sturdy holdfast, the stipe of this alga branches repeatedly and somewhat irregularly. Near the base of the plant, the stipe is cylindrical (to 1 cm in diameter). In the upper portions, it is quite flattened (3 to 5 mm thick) and somewhat broad (10 to 15 cm wide) and bladelike near the tip. The fronds of *Egregia* may be up to 10 m long. The stipe has a fairly uniform width near the base of the plant and a rough surface covered with many blunt projections. From each edge of this stipe, a number of small blades (1 to 2 cm wide) and many small oblong floats are produced. Near the upper end of the plant, there is a transition zone beyond which the tip of the stipe becomes flatter and wider (10 to 15 cm) and has an irregularly grooved surface with marginal blades, but no floats. *Egregia* is usually dark brown to olive green in color, except near the tip where the blade may be lighter and more golden brown in color.
Habitat and Distribution: In moderately exposed waters, *Egregia* is found on rocks in the lower intertidal and subtidal zones, while on more exposed shores it may be found slightly higher up in the intertidal region. This species occurs from British Columbia to California.
Comments: Like many of the other large kelps, *Egregia* has been used as a

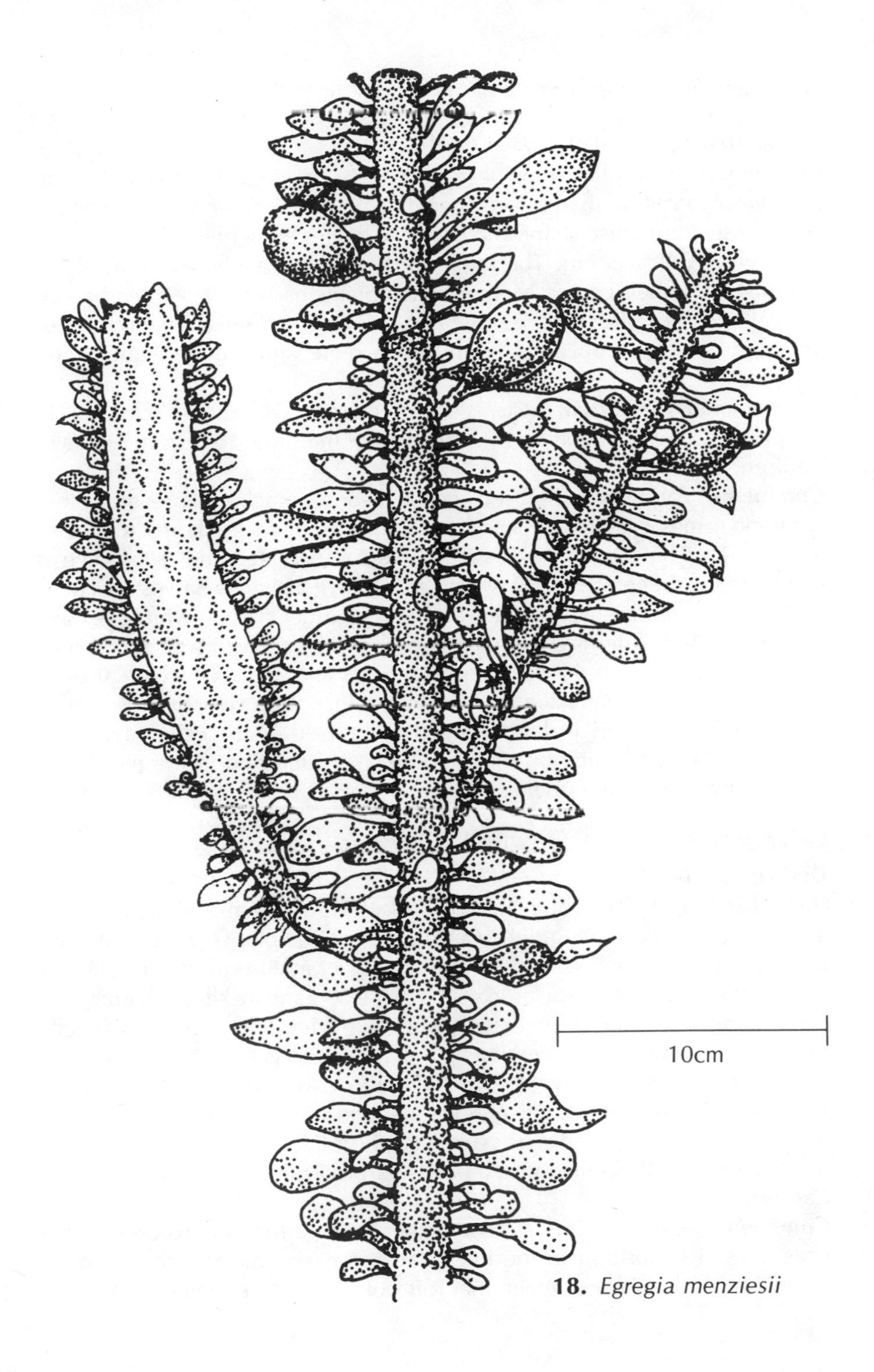

18. *Egregia menziesii*

mulch and fertilizer by coastal farmers.

Fucus distichus Linnaeus *(Pl. 10)*

Description: *Fucus* plants may be from 10 to 50 cm long. When wet, they are olive green to yellowish in color; when dry, they are sometimes almost black. *Fucus* arises from a disc-shaped holdfast and begins branching repeatedly and equally every 2 to 5 cm. The thick but flattened branches have a midrib running down their center. Mature, fertile plants have very swollen branch tips called receptacles. On these swollen branch tips, there are minute pits or depressions called conceptacles in which male and female gametes are produced.

Habitat and Distribution: *Fucus* is one of the most widespread seaweeds in temperate waters. It occurs on rocks in the mid-intertidal zone from Alaska to California.

Comments: Variously called rockweed, bladder-wrack, and several other common names, species of *Fucus* are common along rocky shores in temperate seas. They have been used as mulch, as cattle fodder, and as a source of alginate and other products for which large brown algae are harvested. *Fucus* is also the first representative of the brown algal order Fucales which we have mentioned. This order is a group of highly evolved and quite distinct brown algae. Its members reproduce by shedding eggs and sperm which are usually released when the plants are washed by the incoming tide. Because *Fucus* eggs can be obtained in large quantity, fertilized *Fucus* eggs have been favorite subjects of study by marine biologists interested in the process of development in marine organisms.

Pelvetiopsis limitata (Setchell) Gardner *(Fig. 19)*

Description: *Pelvetiopsis* resembles a miniature *Fucus,* since it has swollen *Fucus*-like branch tips. The plant body arises from a small discoid holdfast, from which one or more upright axes may arise. The plant is repeatedly and usually equally branched, and the branch tips of mature plants are swollen and covered with conceptacles in which the sex organs are situated. Branches of *Pelvetiopsis* are much thinner (only 4 or 5 mm thick) than branches of *Fucus* and have no midrib. The overall length of the plants is typically 8 to 18 cm. Their color ranges from olive green to light yellowish and even reddish brown.

Habitat and Distribution: *Pelvetiopsis* is found only very high in the rocky intertidal zone on exposed coasts subject to the full force of waves spawned in the open ocean. It occurs in such habitats from British Columbia to central California.

Comments: *Pelvetiopsis* is a good indicator organism of exposed rocky coasts. Its high position in the intertidal region and the fact that it forms extensive bands or zones mean that it is not likely to be overlooked.

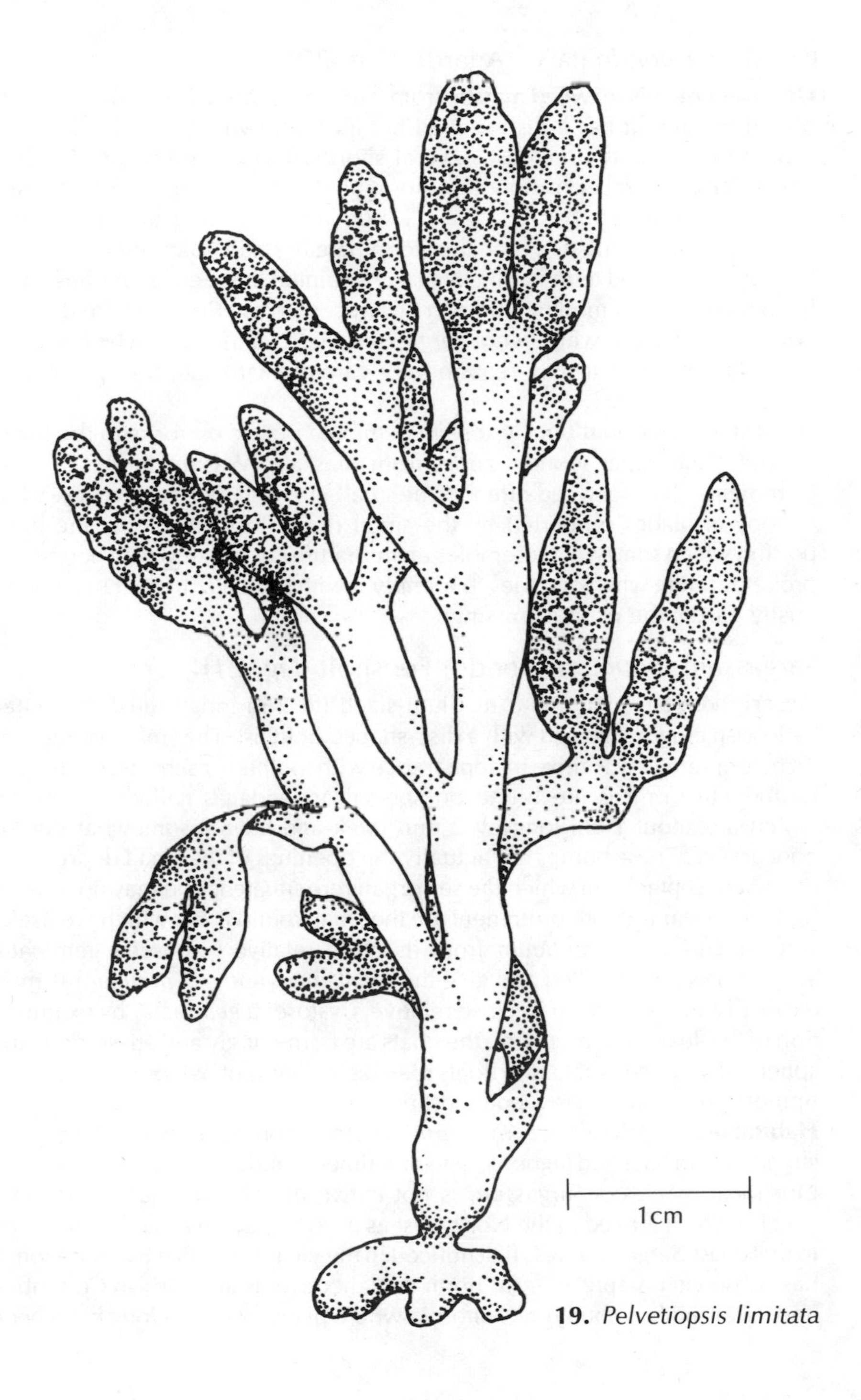

19. *Pelvetiopsis limitata*

Cystoseira geminata C. Agardh *(Fig. 20)*

Description: This seaweed may be from 2 to 5 m long and is yellowish to dark brown in color. It has a disc-shaped holdfast from which a stiff cylindrical stipe arises. Near the holdfast, several short (10 to 15 cm) branches occur which have a number of smooth oblong blades. Farther up the plant, these branches produce more branches which are very long and repeatedly branched. These long thin branches terminate in small floats, often arranged in short chains, and terminating in sharply pointed tips. Sex organs are borne in conceptacles on modified branch tips (receptacles). Compare the description of *C. geminata* with that of *Sargassum muticum (fig. 21)* which follows immediately, since it may require very careful examination to distinguish these two seaweeds!

Habitat and Distribution: *Cystoseira geminata* occurs on rocks in the lower intertidal and upper subtidal zones from Alaska to Washington.

Comments: This seaweed often forms small groves which have a spreading canopy of blades supported by the small floats. In winter, only the basal portion which somewhat resembles an immature *Fucus* frond may persist and prove puzzling when one tries to identify it without realizing that most of the bushy part of the plant is missing.

Sargassum muticum (Yendo) Fensholt *(Fig. 21)*

Description: *Sargassum* is a medium-sized (to 2 m long), much-branched yellowish brown seaweed with a disc-shaped holdfast. The small blades (1 to 2 cm long) are very leaflike in appearance with toothed or serrated edges. The reproductive organs are borne on special appendages called receptacles which are about 1 cm long by 2 mm thick and have a somewhat bumpy appearance. These bumps are actually the openings of the small depressions called conceptacles in which the sex organs are situated. You may not always find fertile plants. More prominent are the small round floats which are useful in distinguishing *S. muticum* from its close relative *Cystoseira geminata.* These floats are gas-filled and give the plants buoyancy. *S. muticum* is most easily distinguished from its close relative, *Cystoseira geminata,* by examination of the floats. In *S. muticum,* the floats are borne singly and are smooth and spherical; in *C. geminata,* the floats may be in chains of two or three, and the tipmost float always has a pointed tip.

Habitat and Distribution: *S. muticum* may form conspicuous stands in quiet, slightly warm bays and lagoons. It is sometimes found cast ashore on beaches. Our local species of *Sargassum* is not native but apparently is a Japanese species which arrived in the Northwest as a stowaway on oysters introduced to this coast. *Sargassum* was first noticed in the vicinity of Coos Bay, Oregon. It has subsequently spread as far north as Vancouver Island, British Columbia, and as far south as southern California where plants over 3 m long have been

PLATE 1

Egregia menziesii

Platymonas sp.

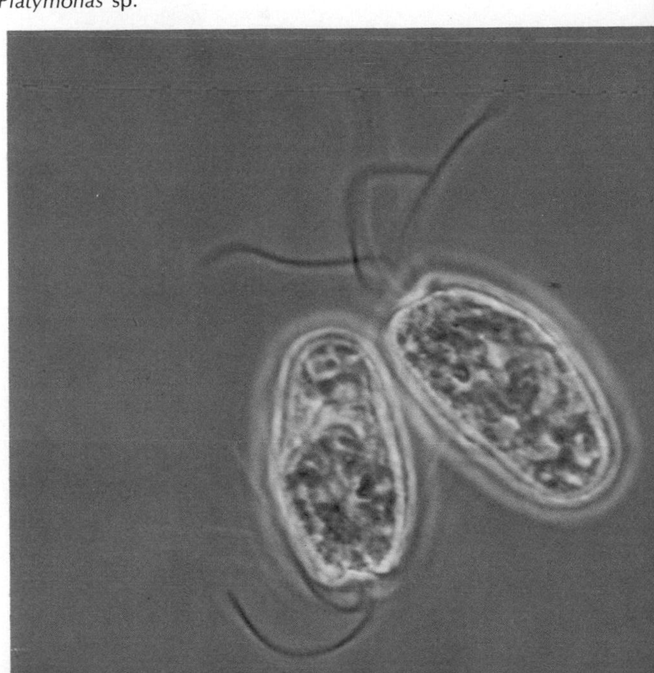

Enteromorpha prolifera

Monostroma arcticum

Ulva lactuca

Urospora mirabilis

Cladophora trichotoma

Acrosiphonia coalita

Halicystis ovalis

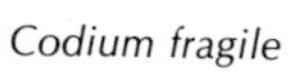

Codium fragile

PLATE 4

Colpomenia bullosus

Scytosiphon lomentaria

Leathesia difformis

PLATE 5

PLATE 6

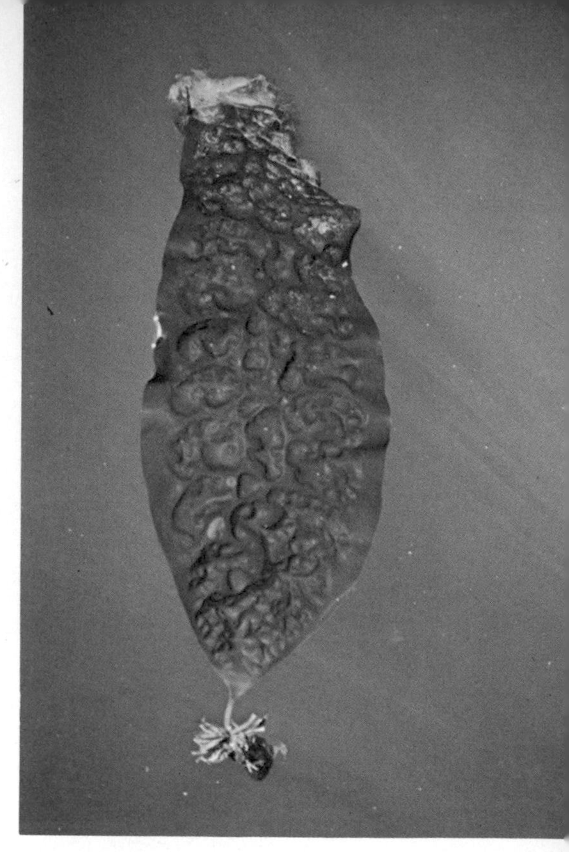

Laminaria saccharina

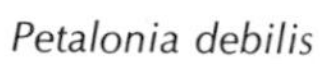

Petalonia debilis

Laminaria groenlandica

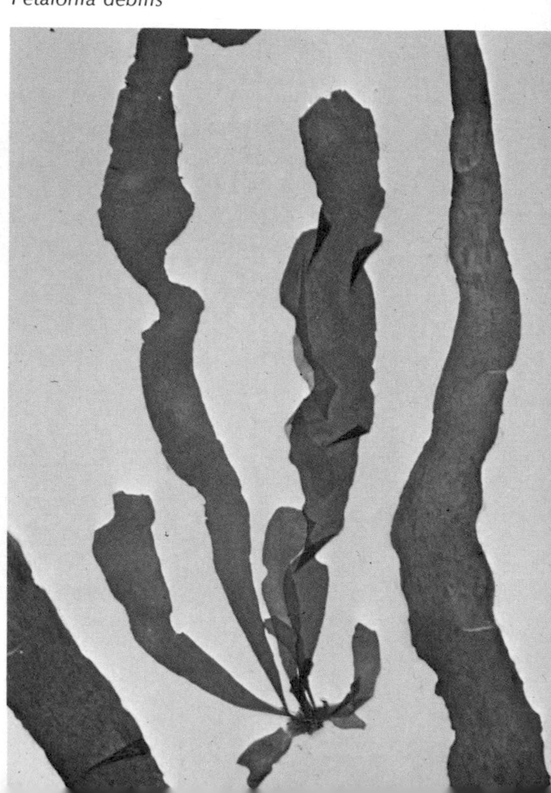

Agarum cribrosum

Hedophyllum sessile

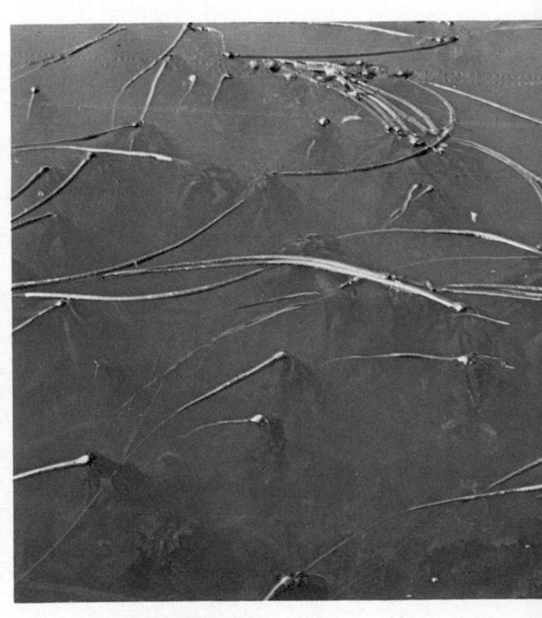

Nereocystis luetkeana

PLATE 8

Postelsia palmaeformis

Macrocystis integrifolia

PLATE 9

Alaria marginata

Pterygophora californica

Lessoniopsis littoralis and *Lithothamnium pacificum*

PLATE 10

Gelidium robustum

Smithora naiadum

Fucus distichus

PLATE 11

Corallina vancouveriensis

Plocamium coccineum

PLATE 12

Iridaea cordata

Iridaea cornucopiae

Palmaria palmata

PLATE 13

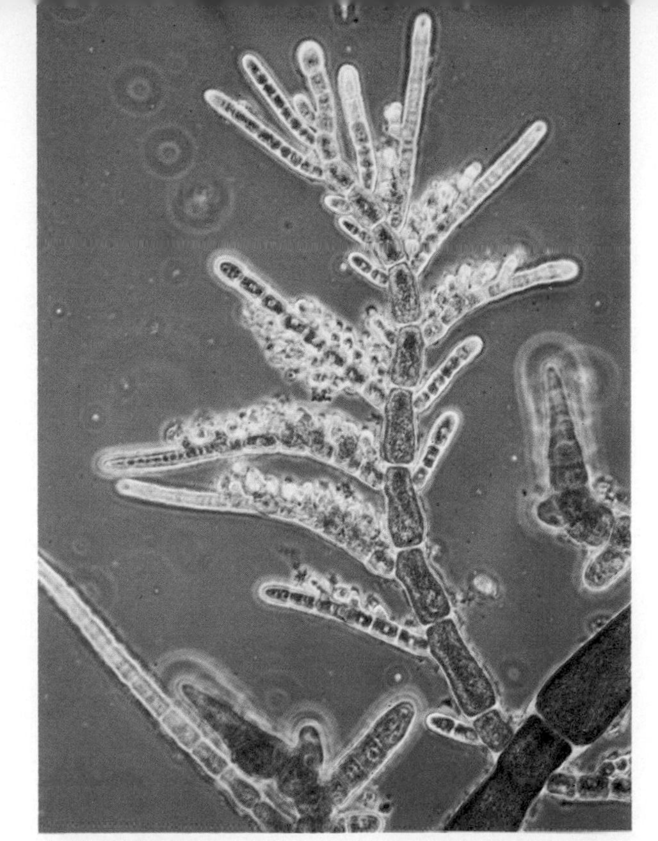

Antithamnion sp. *Ceramium gardneri*

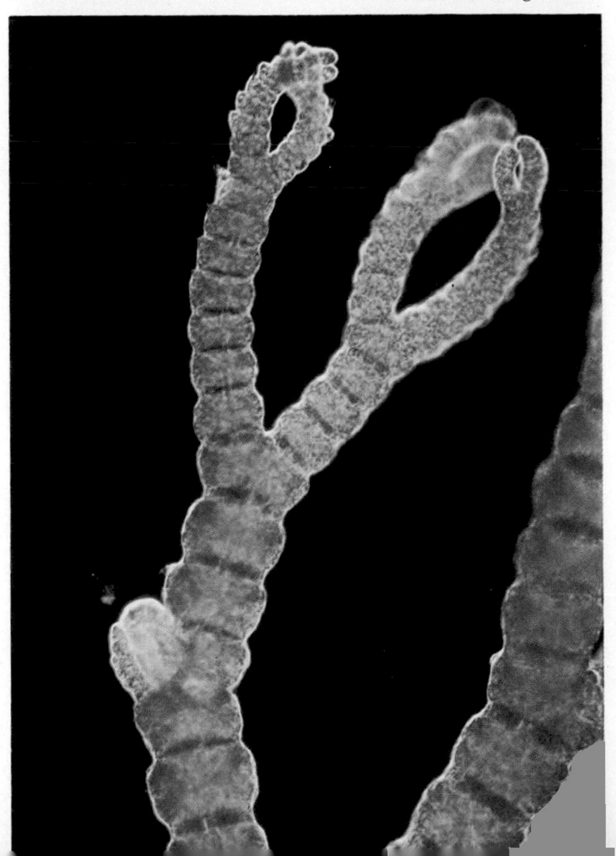

PLATE 14

Delesseria decipiens

Botryoglossum farlowianum

Rhodymenia pertusa

PLATE 15

Odonthalia floccosa

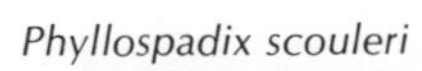

Phyllospadix scouleri

PLATE 16

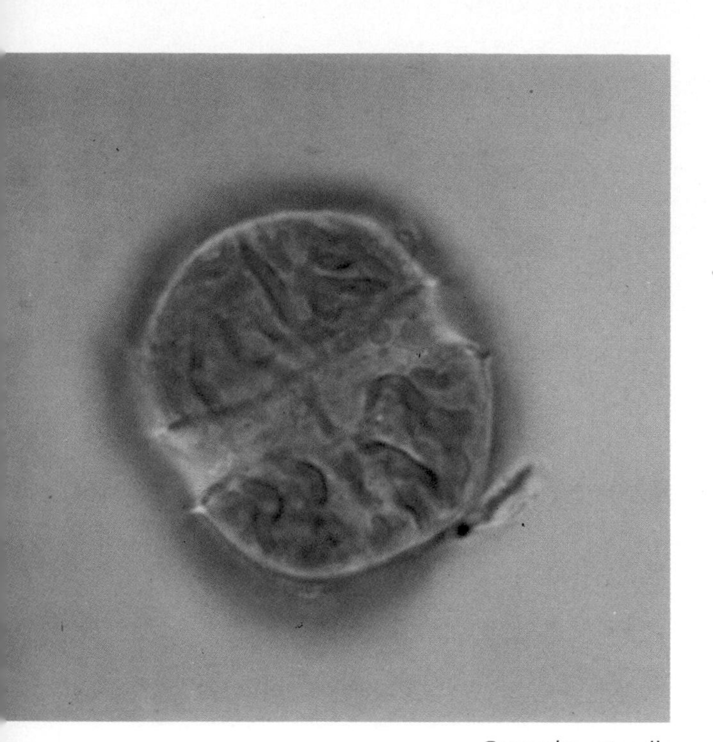

Gonyaulax catenella

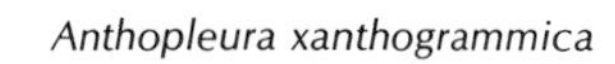

Anthopleura xanthogrammica

20. *Cystoseira geminata*

found. *S. muticum* is evidently quite peripatetic, for in 1973 it was found in Great Britain.

Comments: The floats which render it buoyant are probably the most important feature of *Sargassum*. These contribute to its widespread occurrence and its success as an invader of new territory. When a branch breaks loose from the main plant body, it may hitch a ride on a water current and be rafted to some other area. While most *Sargassum* species are prominent in warmer tropical seas, *S. muticum* must have a wide tolerance for water temperatures or be able to adapt quickly in order to successfully colonize such a large stretch of new territory as it has done on our Pacific Coast. Perhaps the most famous *Sargassum* is *S. natans,* which forms extensive free-floating masses in that region of the North Atlantic called the Sargasso Sea. In this area, a great confluence of ocean currents forms a one-way trap for floating debris. The population of *Sargassum* there is believed to have been started by the arrival of floating material from some rocky Atlantic shore. The present bulk of the population is believed to be maintained exclusively by vegetative reproduction. The Sargasso Sea is an example of a rare situation in which a really large seaweed has a truly pelagic existence. Masses of the highly branched floating plant bodies provide shelter for a unique community of animals found nowhere else in the world. The development of such a unique community suggests that the Sargasso Sea has been in existence for a very long time.

Red Algae (Rhodophyta)

This group of algae has a number of representatives in fresh water but reaches its greatest abundance and diversity in marine waters. Plants in this group contain the green pigment, chlorophyll, whose presence is masked by an abundance of red and blue pigments responsible for the overall color of the plants. A kind of starch, called floridean starch, which is identical to the starch from green land plants, is the form in which the photosynthetically produced food reserves are stored. The cell walls of these algae contain cellulose and other polymers, such as agar and carrageenan for which certain species are harvested commercially.

Although the simplest red algae are tiny single cells (only visible in the microscope), most are easily visible to the naked eye. Some are 2 to 3 m long. These larger forms may be composed of sheets of cells or elaborately interwoven filaments of cells. Some red algal cells can creep about in ameboid fashion, but none are able to swim about rapidly, since no red algae have flagella.

There are about 115 genera and more than 265 species of red algae in the Pacific Northwest. Many of these are quite difficult to identify accurately; many are quite rare or inconspicuous. Therefore, I have described only the most common genera and species in the following pages.

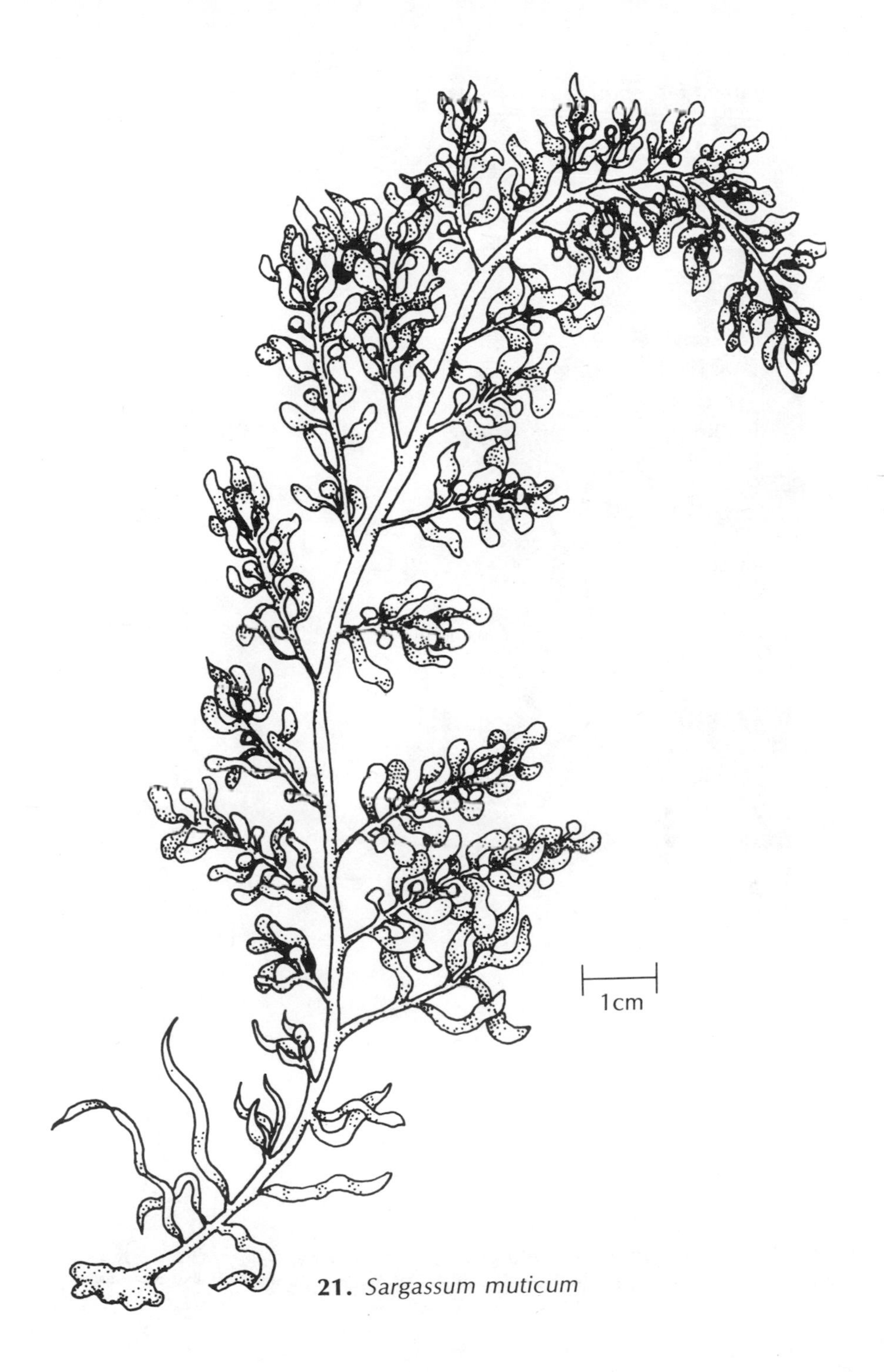

21. *Sargassum muticum*

22. *Bangia fuscopurpurea* -Habit

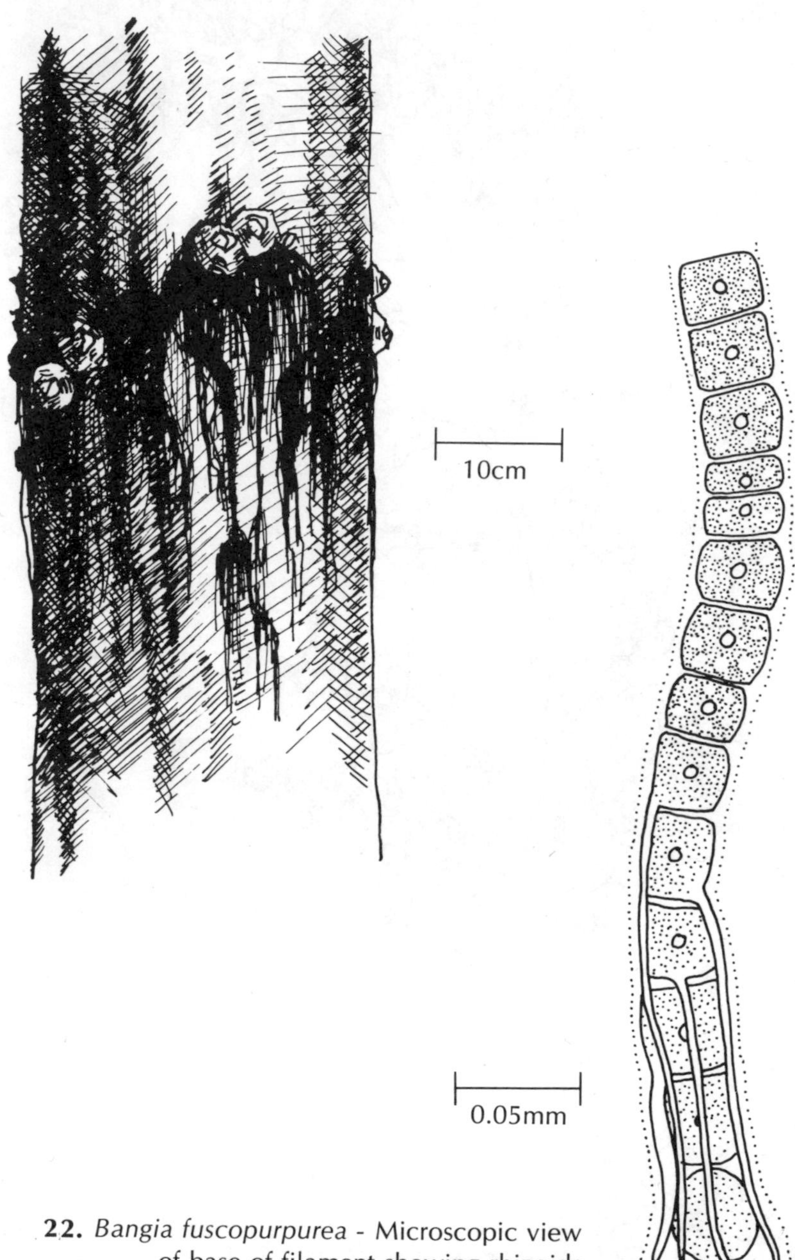

22. *Bangia fuscopurpurea* - Microscopic view of base of filament showing rhizoids

Bangia fuscopurpurea (Dillwyn) Lyngbye *(Fig. 22)*

Description: This alga frequently appears as dark purple red to black streaks on pilings and rocks in the upper intertidal zone. The streaks are composed of groups of individual, cylindrical filaments from 2 to 10 cm long. Near their bases, these filaments consist of only a single row of cells, while near the tip they are usually several cells thick. Cells near the base of a filament have long rhizoidal extensions which attach the individual plants to the substrate. A good hand lens or low-powered microscope will be of immense aid in examining *Bangia*.

Habitat and Distribution: *Bangia* is often most noticeable high in the intertidal zone on rocks, pilings, and old wood. It is found in both exposed and protected habitats from Alaska to Central America.

Comments: *Bangia fuscopurpurea* is an excellent example of a species which has probably developed local races or strains called ecotypes. Ecotypes are local populations of a species that are adapted to local water conditions, especially temperature. In the case of *B. fuscopurpurea,* specimens which appear to belong to the same species can be found in near-freezing Alaskan waters and very warm tropical waters. There are only a limited number of marine algal species which are distributed over such a great range of sea temperatures.

Porphyra perforata J. Agardh *(Fig. 23)*

Description: Many species of *Porphyra* occur in the Pacific Northwest, but only one of these *P. perforata,* or purple laver, will be discussed here. It is a thin blade which is actually only one cell thick (some species of *Porphyra* are two cells thick) and is attached to rocks or other objects by means of a tiny discoid holdfast. When thoroughly wet, the plants appear steel grey to dull purple in color. When exposed to the air for several hours (as on a low tide), they may appear nearly black and have a very stiff, rubbery, or even dry and brittle texture. Individual blades may be deeply lobed and have ruffled edges.

Habitat and Distribution: *P. perforata* can be found from Alaska to Mexico and grows on rocks in the mid- to high intertidal zone. Some species grow attached to other seaweeds, and some species grow only subtidally.

Comments: *Porphyra* is one of the better known seaweeds, since it is eaten by many people and is the heart of a multimillion dollar business which employs thousands of people. It has been called *laver* (water plant) by peoples of European descent and is known as *nori* in the Orient. An ingredient in a variety of culinary delights, *Porphyra* is used in soups and stews, as well as in many rice dishes. Details of its cultivation are given in Chapter V.

Smithora naiadum (Anderson) Hollenberg *(Pl. 10)*

Description: This small purple red blade (2 to 10 cm long) grows on the

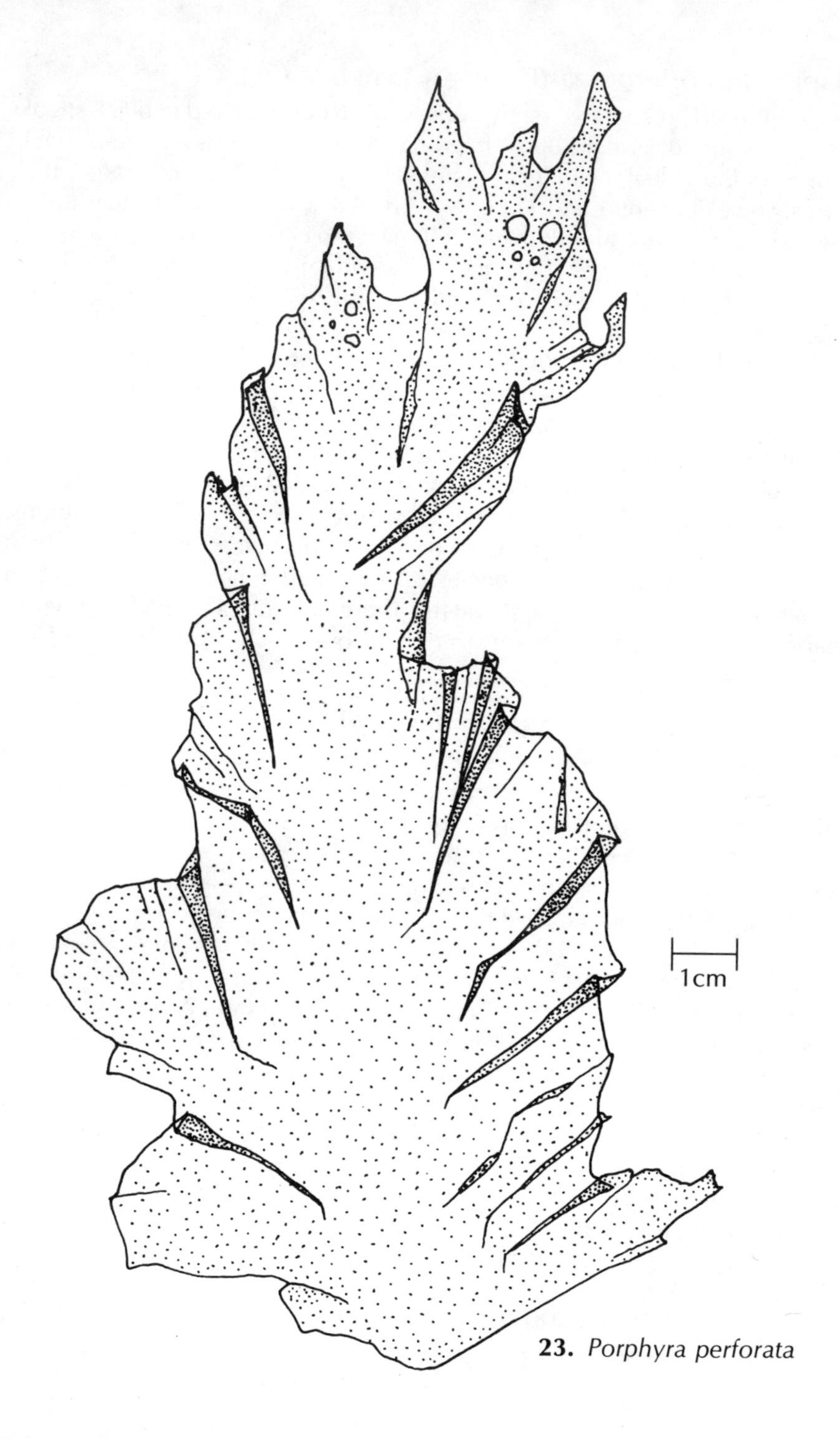

23. *Porphyra perforata*

surface of the long strap-shaped leaves of eelgrass, *Zostera,* and surfgrass, *Phyllospadix*. When matted or clumped together, *Smithora* blades appear dark purple black, but their deep red purple color is best seen if they are viewed against a light background.

Habitat and Distribution: Since *Smithora* is an epiphyte which is only found growing on *Zostera* or *Phyllospadix,* its distribution goes hand in hand with the distribution of *Zostera* (often abundant in the upper subtidal region of quiet bays with mud bottoms), and *Phyllospadix* (occurring on rocks in the lower intertidal and upper subtidal zones). *Smithora* and both of these sea grasses are found from Alaska to Mexico on the Pacific Coast. *Smithora* is most abundant in the late spring and summer.

Comments: Once considered a species of *Porphyra, Smithora* was described as a separate genus when laboratory studies revealed that it has a different pigment composition and a different pattern of reproduction and early development than is found in *Porphyra*.

Bonnemaisonia nootkana (Esper) Silva *(Fig. 24)*

Description: Plants of *Bonnemaisonia* are usually bright red in color and may be up to 45 cm tall. They may have more than one main branch. The branches are cylindrical or slightly flattened (up to 3 mm in diameter) and have two rows of opposite branches. Of each pair of opposite branches, one is usually short and simple, while the other is long and repeatedly branched. It often has inflated hooklike branches at irregular intervals in place of these longer branches. The tips of the lateral branches are usually sharp and pointed.

Habitat and Distribution: *Bonnemaisonia* grows attached in the very low intertidal and subtidal regions from northern British Columbia to Baja California.

Comments: The macroscopic *Bonnemaisonia* plants are the sexual phase of this alga—an example of an alga with alternating heteromorphic generations. There is also a free-living microscopic, filamentous, spore-producing phase which was named *Trailliella* before it was known to be part of the life history of *Bonnemaisonia*. At first glance, it is easy to confuse *B. nootkana* with another red alga, *Ptilota filicina*.

Gelidium robustum (Gardner) Hollenberg and Abbott *(Pl. 10)*

Description: *Gelidium* occurs in tufts or clumps from 6 to 25 cm high. The plants are quite stiff and may be from dark purple red to somewhat brownish red in color. The axes of the branch systems are flattened and up to 3 mm wide. A succession of branches, arranged in two opposite rows, are produced near the tips of the principal axes and primary branches. Near the tip of each axis, young branches are parallel to the axis, but farther back from the tip the branches lie at right angles to the principal axes. The plant attaches to the

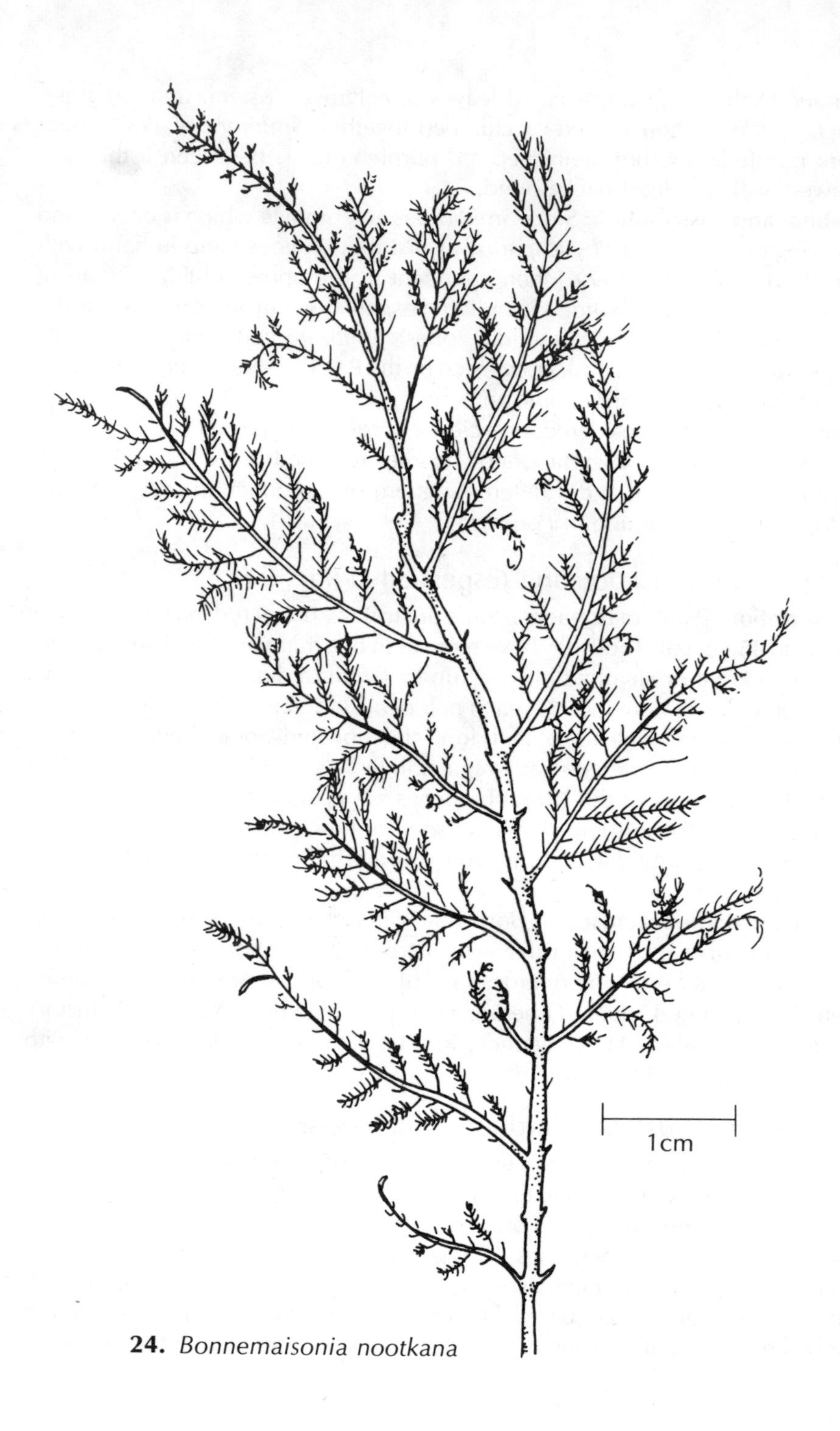

24. *Bonnemaisonia nootkana*

substrate by a system of prostrate branches.
Habitat and Distribution: *G. robustum* occurs on rocks in the upper subtidal and lower intertidal zone from British Columbia to Baja California. It never grows very large in the Northwest; however, in warmer waters it reaches a much larger size. There are several other species of *Gelidium* of differing size, color, and branching pattern which you might also find in Pacific Northwest waters.
Comments: In other parts of the world where it is more abundant, *Gelidium* is harvested as a source of agar-agar which is used in a variety of industrial products, such as microbial culture media, prepared foods, and textile manufacturing.

Corallina vancouveriensis Yendo *(Pl. 11)*

Description: Bright pink and very tough, *Corallina* and other jointed, or articulated, coralline algae are often seen in the intertidal and subtidal regions. *Corallina* consists of a crustose base from which many erect, branched, jointed, and flexible axes arise. The upright portion is 4 to 10 cm high. The branch segments are cylindrical to slightly flattened, 0.7 to 1 mm in diameter. Most of the lateral branches are arranged in two opposite rows; they are all about the same length and very close together which gives the plant a broad, flattened appearance. Occasionally, the branching pattern is more complex.
Habitat and Distribution: A frequent inhabitant of tide pools, *Corallina* is widely distributed on rocks in the lower intertidal and upper subtidal zones. It is found from Alaska to Mexico.
Comments: The heavily calcified, articulated coralline algae were once thought to be animals closely related to corals, since their cell walls contain massive deposits of calcium carbonate. At least two more species of *Corallina* are also found in local waters, along with several more genera of articulated corallines, such as *Calliarthron, Bossiella,* and *Serraticardia,* which are also quite common in the Northwest. Bleached specimens of these algae turn pale white and are often found washed ashore.

Lithothamnium pacificum Foslie *(Pl. 9)*

Description: *Lithothamnium* appears as bright pink, often very extensive, calcareous crusts growing on rocks, shells, and other substrates. The edge of such a crust is circular or irregular in outline; several crusts may grow together and overlap. Small holes in raised protuberances mark the openings from which spores may be shed.
Habitat and Distribution: *Lithothamnium* and related genera are quite common in tide pools, in the lower intertidal and subtidal zones on rocky surfaces. *L. pacificum* is found from British Columbia to California.
Comments: Although other encrusting species of coralline algae are common

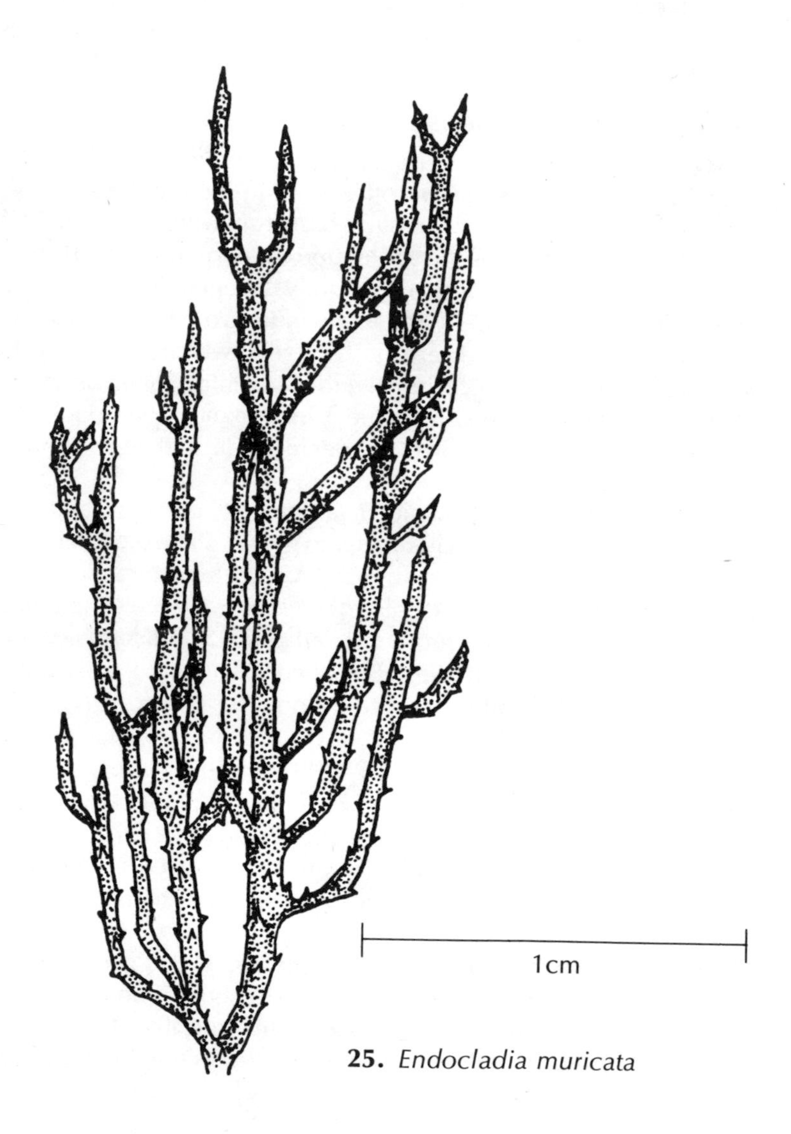

25. *Endocladia muricata*

inhabitants of all seas, difficulties inherent in collecting them and working with them have resulted in a general neglect of these algae by most marine botanists. They are usually among the deepest growing of all marine algae. Because generation after generation of encrusting coralline algae leaves thick deposits of calcium carbonate, these algae are very important as reef builders in tropical waters.

Endocladia muricata (Harvey) J. Agardh *(Fig. 25)*

Description: This alga forms dark brownish red or purple to black clumps 4 to 8 cm tall. The clumps are often associated with barnacles. When dry, *Endocladia* is very wirelike and hard; when moist, the plant is softer and more flexible but still retains an upright stature and wiry appearance. The cylindrical branches (0.5 to 1 mm in diameter) are covered with short spines (.05 mm), and the plant is profusely branched.

Habitat and Distribution: On moderately to very exposed rocky shores, *Endocladia* and the small barnacle, *Balanus,* form a distinct association high in the intertidal region. This association is a conspicuous feature of rocky shores from Alaska to Mexico.

Comments: An intensive study of the *Endocladia-Balanus* community has revealed that organisms living at this level of the intertidal zone are exposed to air longer (73 percent of the time) than they are submerged (only 27 percent of the time). Plants and animals at this level of the intertidal zone may be exposed to air for as many as twenty-four days at a time; however, they will certainly be moistened by wave splash or spray during such periods of exposure. Numerous other small algae and animals (more than 90 species) live in the crevices and shelter of this plant and animal association.

Prionitis lanceolata Harvey *(Fig. 26)*

Description: This alga is typically 10 to 35 cm tall and may be dull, brownish red to dark purple in color. It is attached to the substrate by a small discoid holdfast from which a blade arises. The blade, narrow near the bottom, soon broadens out into a flattened shape 3 to 8 mm wide and usually terminates in a point. This main blade typically has numerous, oppositely arranged, lateral blades of a similar shape which are slightly smaller than the main blade. The large number of lateral blades gives this plant a somewhat clumped appearance.

Habitat and Distribution: This alga (and some related species) occurs from Alaska to Mexico. It is especially prominent in tide pools where it often has the dull reddish brown color mentioned above. It can also be found in the lower intertidal and subtidal zones where it is usually dark red or purple.

Comments: If one carefully smells a clump of *Prionitis,* a definite iodine odor can be detected; it smells very much like the seashore on a foggy day.

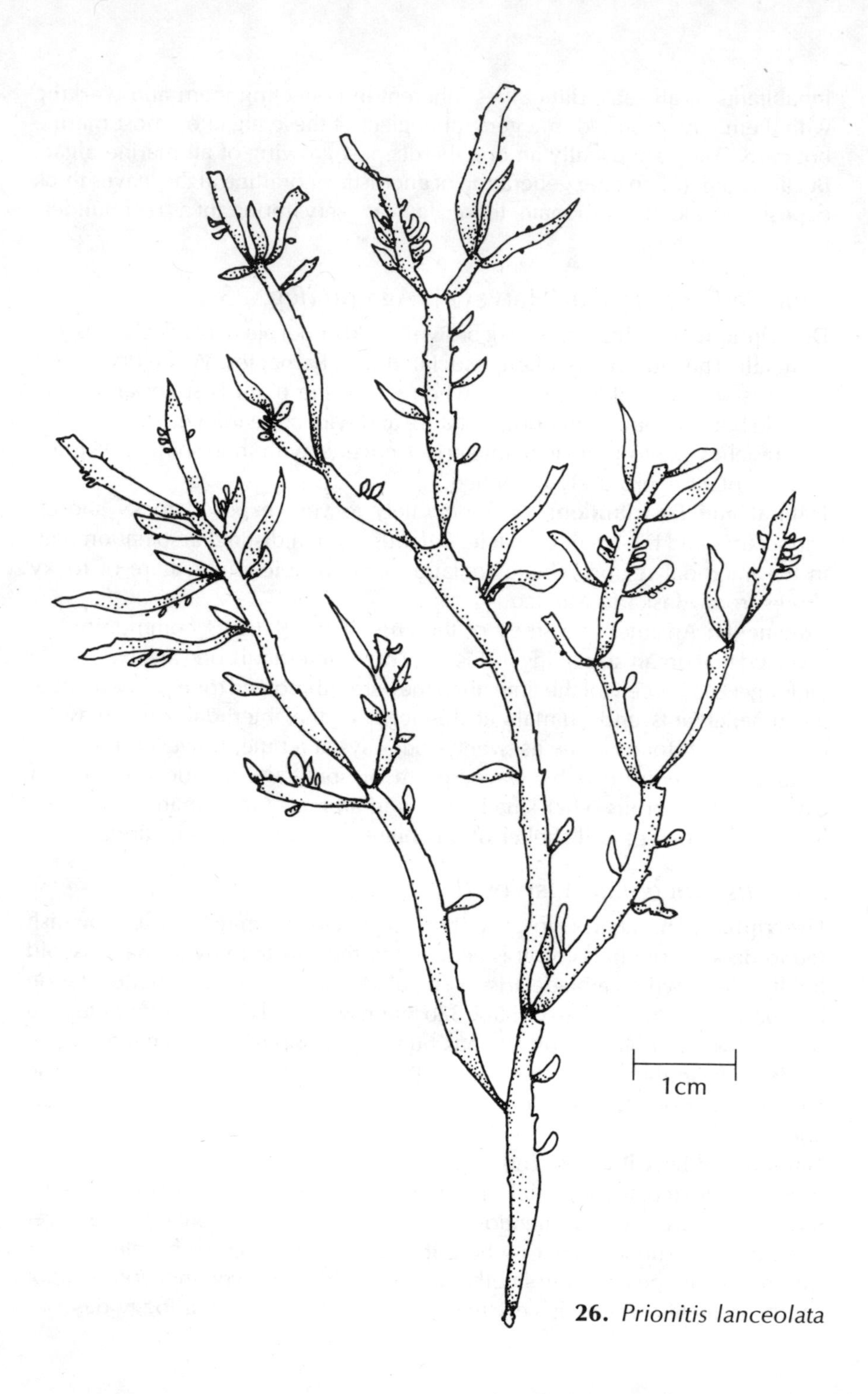

26. *Prionitis lanceolata*

Callophyllis flabellulata Harvey *(Cover photo)*

Description: This deep red alga has a fan-shaped appearance. It arises from a small discoid holdfast and has a narrow flat stipe. The stipe divides repeatedly in one plane to produce broad but thin segments (1 to 2 cm wide) which are then repeatedly divided into narrower segments near the tips of the blades. The overall height of these plants is from 15 to 20 cm.
Habitat and Distribution: This species and its close relatives are usually found from the very low intertidal to the subtidal regions. It frequently washes up on the beach. This species occurs from Alaska to California.
Comments: It is quite difficult to correctly identify some species of *Callophyllis,* requiring material that is in reproductive condition to make critical determinations. These criteria are frequently encountered in distinguishing between certain red algae which superficially resemble one another. There are many other species of red seaweeds which are more or less fan-shaped.

Neoagardhiella baileyi (Harvey ex Kützing) Wynne and Taylor *(Fig. 27)*

Description: *Neoagardhiella* plants typically measure from 15 to 40 cm high and may grow even larger. They have a minute, discoid holdfast which is usually attached to a small pebble or piece of shell. From the holdfast, 2 to 3 mm in diameter, the main axis arises. This axis is often obscured by a profusion of branches of nearly the same diameter and length as the main axis. *Neoagardhiella* is brownish to deep red in color, with a soft, almost slimy texture. Both the main axis and the branches are narrower at their bases than near their midsections. The branches may be arranged radially, so that the plant appears bushy, or they may be nearly all in the same plane, so that the plant appears flattened.
Habitat and Distribution: *N. baileyi* is found from Alaska to Mexico. In the Northwest, it is quite common in areas with relatively sluggish currents, and with sand, gravel, or shell-strewn bottoms. It also occurs in more rocky areas. It may be found growing attached in the very low intertidal and subtidal zones from 10 to 15 mm deep. It grows attached to small pebbles in *Zostera* beds and similar places and is commonly found cast up on beaches.
Comments: When fresh, *Neoagardhiella baileyi* (formerly known as *Agardhiella tenera)* may be used as a condiment in salads. In the Philippines, it is boiled with sugar and spices into a sweet. There are two other algae in this area which resemble *Neoagardhiella* and often occupy similar habitats. These are *Gracilariopsis sjoestedtii* and *Gracilaria verrucosa,* which are usually darker in color, with thinner and fewer branches than *Neoagardhiella.*

27. *Neoagardhiella baileyi*

Plocamium coccineum var. *pacificum* (Kylin) Dawson *(Pl. 11)*

Description: Plants of this species are bright to dark red in color and profusely branched. They may be from 10 to 30 cm high, somewhat bushy in appearance or slightly flat in form. A small discoid holdfast attaches the plant to the substrate. The stipe, and the branches arising from the stipe, are all somewhat flattened and 1 to 2 mm wide. Careful examination will show that all the smaller branchlets are borne on one side of the larger branches. This one-sided branching pattern occurs even in the smallest branches which are scimitar-shaped and bear branches on the side toward the axis on which they are borne. This branching pattern is an excellent aid in recognizing *Plocamium*.

Habitat and Distribution: *Plocamium* is found on rocks from the lower intertidal area to depths of 10 to 15 m in the subtidal zone. This species is found from British Columbia to Mexico. There are several other species of *Plocamium* found in Northwest waters, but none are as common as *P. coccineum*.

Comments: A careful examination of *Plocamium coccineum* plants will often reveal tiny (1 to 2 mm) cushions of spiny-appearing white or brownish white tissue on its branches. This mass of tissue is actually a species of red alga that has apparently lost its ability to form photosynthetic pigments and perform photosynthesis; hence, it is regarded as a parasite on the host plant. This particular species of parasite is called *Plocamiocolax pulvinata*. There are a number of other examples of such parasitic red algae; their characteristic structure and reproductive organs are the main clue to identifying them as "colorless" red algae. Many of these parasitic red algae are believed to be quite closely related to their host plants; they are quite specific in only parasitizing certain host species.

Petrocelis middendorffii (Ruprecht) Kjellman *(Fig. 28)*

Description: This alga has two forms in the intertidal zone. One of these forms is a dark purple red to black crust, 1 to 2 mm thick and as much as 1 m in diameter, and is an asexually reproducing plant with many minute upright filaments. These are tightly pressed together and arise from a layer of cells located next to the substrate. The other form is an upright blade, up to 15 cm tall, which is often branched and may be from 0.5 to 10 cm wide. The blade's surface is covered with many papillae; its texture is very stiff and rubbery. Such blades may vary in color from yellowish brown to dark purple red and even black. Known for many years as *Gigartina papillata,* the dark blades are now believed to be the sexually reproducing phase of *Petrocelis*.

Habitat and Distribution: The encrusting form of this plant is perennial and very long-lived. It is found in the mid-intertidal zone where it is often abundant on rocky shores. The upright, blade-forming phase of this plant is also found in the mid-intertidal zone. Although most abundant in spring and summer, it can be found throughout the year. Both forms are found from Washington to Mexico.

Comments: Only very recently have laboratory and field studies revealed that the encrusting red alga, *Petrocelis middendorffii,* and the upright plants, known as *Gigartina papillata* and *G. agardhii,* are part of the life history of the same species. The crustose plants are asexual, producing spores called tetraspores which germinate to produce the blade forms. The blade forms are either male or female; they reproduce sexually. In the blade form, fertilization results in spores called carpospores. At least some of the carpospores germinate to produce the crustose form. As a potential source of the commercially useful polymer, carrageenan, it is essential that the details of the life history and pattern of reproduction of species such as this be thoroughly studied. Many of the encrusting red algae have not been very thoroughly investigated, and further surprises may be in store for those who do study them. It seems likely that other plants now considered to be species of *Gigartina* may really be phases in the life history of *Petrocelis* or some other alga. Still other species of *Gigartina,* such as *G. exasperata,* seem to have both asexual and sexual plants that form large blades.

28. *Petrocelis middendorffii* - Habit

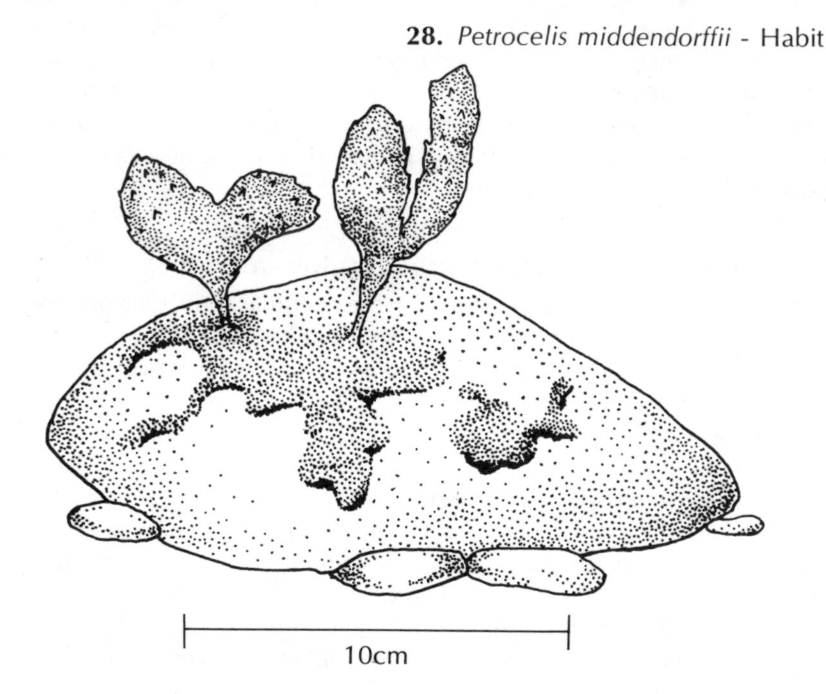

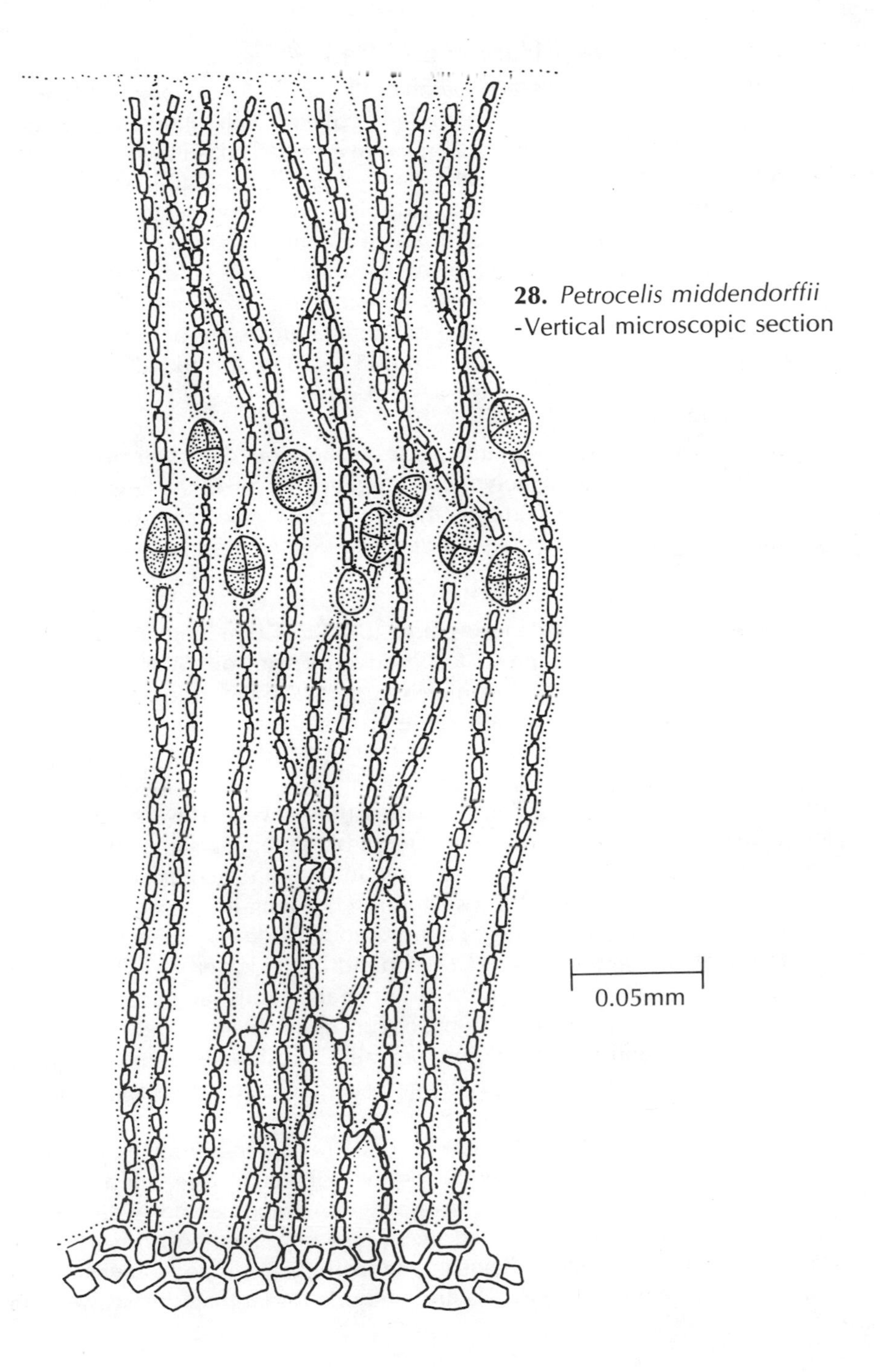

28. *Petrocelis middendorffii*
-Vertical microscopic section

Gigartina exasperata Harvey and Bailey *(Fig. 29)*

Description: This plant has large broad blades, 30 to 50 cm long and 10 to 20 cm wide or larger. The blades are densely covered with numerous protuberances or papillae which give the appearance of a "Turkish towel"—a common name for this alga. The blades range from deep red to purplish red in color. When submerged, they may exhibit a bluish iridescence. The small stipe is sometimes branched and may give rise to more than one blade. The plant is attached to the substrate by means of a discoid holdfast which may vary from a few millimeters to several centimeters in diameter.

Habitat and Distribution: *Gigartina exasperata* grows on rocks in the lower intertidal and subtidal zones at depths of 5 to 10 m. It is found from British Columbia to California. There are several other species of *Gigartina* in the Pacific Northwest.

Comments: A number of the many species of the genus *Gigartina* are valuable as producers of the marine polymer, carrageenan. One of the larger of these species, *G. exasperata,* is currently the subject of a research project investigating the feasibility of farming this species. (See Chapter V).

Iridaea cordata (Turner) Bory *(Pl. 12)*

Description: *Iridaea cordata* forms large (0.25 to 1.0 m long), conspicuous, iridescent blades which appear to be covered with oil. When lying submerged, the blades appear dark bluish purple, and the iridescence is quite attractive. The blades are usually quite broad; they may have a smooth or undulating margin. Frequently, they are split or lobed. There is a small stipe at the base of the blade, and the plant is attached to the substrate by an encrusting discoid holdfast which may measure between a few mm to several cm in diameter. Often, a group or cluster of blades appears to grow from the same holdfast, but usually only one blade in a group becomes very large. It now appears that the small crustose holdfast is a long-lived, perennial entity which annually sprouts the large, conspicuous blade.

Habitat and Distribution: *I. cordata* is found in the lower part of the rocky intertidal area where it may form a noticeable zone or band. It is also prominent in the upper subtidal zone, occurring in extensive beds where the proper substrate and water conditions prevail. Occasionally, plants may grow in waters as much as 20 m deep. Along the Pacific Coast, this species occurs from Alaska to California.

Comments: As one of the larger species of carrageenan-producing red algae in the Pacific Northwest, *I. cordata* is the subject of research at several university and government laboratories where methods are being developed to utilize this previously untapped living resource. Methods for surveying, mapping, harvesting, and judiciously managing natural populations of this seaweed are being developed, as are methods for farming this seaweed in

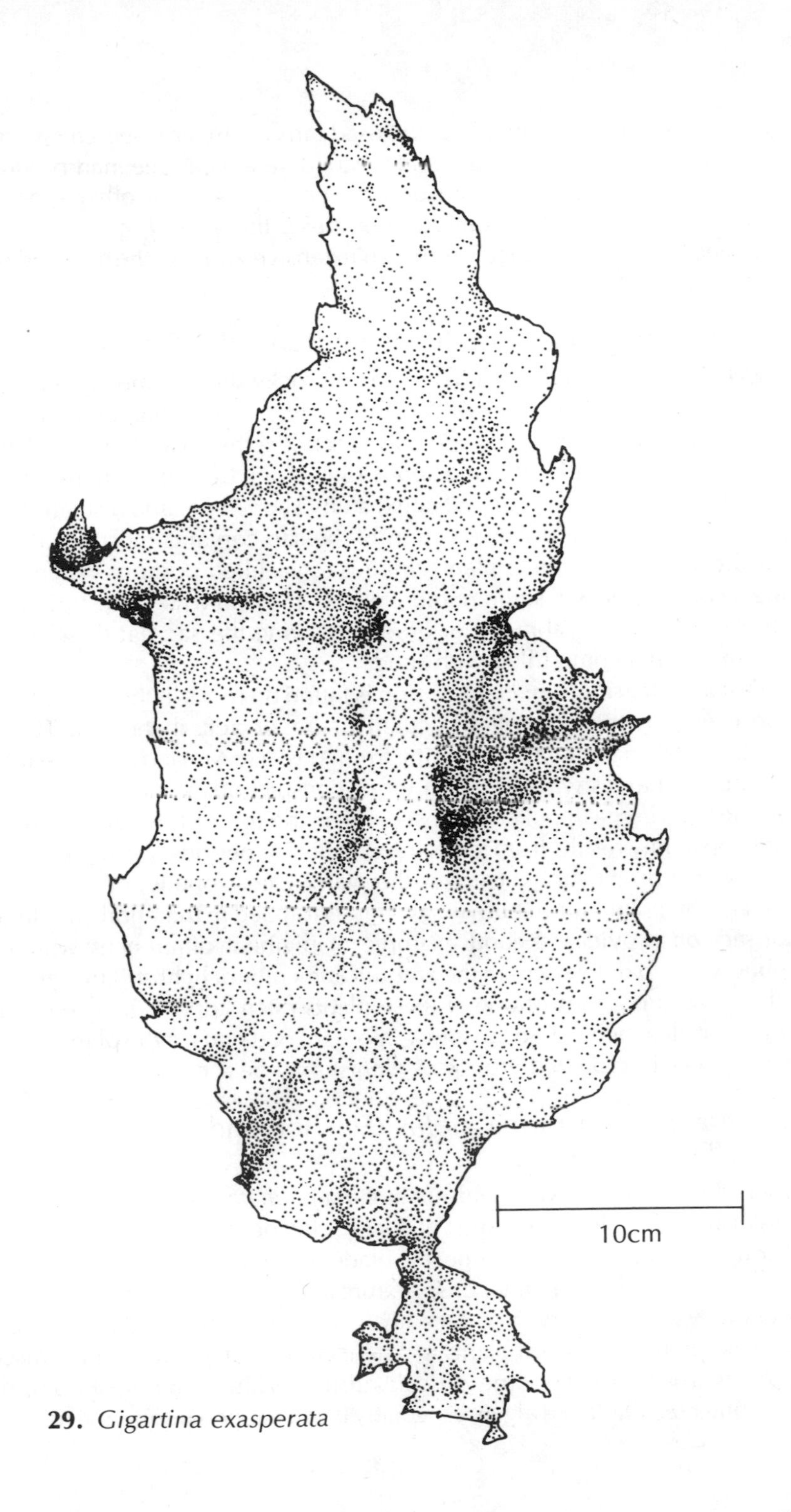

29. *Gigartina exasperata*

both open water farms and large enclosed tanks. Similar research efforts are underway in other parts of the world where different carrageenan-producing species of red algae are being studied. There are several other species of *Iridaea* in regional waters; only one of these, the small *I. cornucopiae,* is illustrated here *(pl. 12),* since it forms an extensive zone on the exposed outer coast (see Chapter III).

Halosaccion glandiforme (Gmelin) Ruprecht *(Fig. 30)*

Description: In the mid-intertidal region on rocky shores, one often encounters clumps of hollow, thin-walled, elongate sausage-shaped sacs. These thalli of *Halosaccion glandiforme* may reach lengths up to 25 cm and 3 to 4 cm in diameter; typical sizes are about 15 cm long by 2 to 3 cm in diameter. Each plant has a broad, rounded apex, a tiny short stipe, and a small discoid holdfast. Usually, the hollow portion of the sac contains mostly water, but near the top in most intact plants there is a gas-filled portion. When gently squeezed, this alga will squirt numerous fine sprays of water. The tips of older plants often become abraded or eroded, leaving the sac flat or sometimes allowing it to become filled with sand.
Habitat and Distribution: *Halosaccion* is found on rocks in the mid-intertidal zone. Clumps of this plant often occur in fissures or clefts in the rock. This alga is widespread on our coast, occurring from Alaska to Mexico in both exposed and sheltered areas where it often forms conspicuous stands.
Comments: The gas pocket often seen near the top of this sausage-shaped plant apparently results from gases trapped during periods of rapid photosynthesis, when bright sunlight strikes the plant. Occasionally, one sees intact *Halosaccion* plants adrift, buoyed by the gas trapped in their hollow interiors. *Halosaccion* provides a good example of pigment differences which are probably due to differences in the amount and color of light striking the plant. Fully exposed portions of one of these plants often appear pale yellow and somewhat bleached, while the lower, more shaded parts of the plant (or small young plants in crowded groups of plants) may be a dark purple red.

Palmaria palmata forma *mollis* (Setchell and Gardner) Guiry *(Pl. 12)*

Description: Also known as dulse, *Palmaria palmata* plants are bright red to brownish red in color and consist of deeply cleft or branched blades up to 50 cm long. Arising from a short stipe, the blade segments may vary in width from 2 to 15 cm. They have a rather crisp texture and consist of a central portion of moderately large cells with surface layers of smaller cells. *P. palmata* often occurs as groups of several plants in a tight clump, attached by small discoid holdfasts. You may have some difficulty distinguishing *P. palmata* from some of the other red bladelike algae, especially if you happen upon small or young plants.

Habitat and Distribution: *Palmaria palmata* is a very cosmopolitan seaweed found in the Atlantic as well as the Pacific. On the Pacific Coast from Alaska to California, *P. palmata* is found on rocky shores. It is fairly common in the lower intertidal zone on both exposed shores and in the more sheltered waters of Puget Sound and Hood Canal.

Comments: Formerly known as *Rhodymenia palmata* and familiar to many people as dulse, *P. palmata* has for centuries been a traditional food item for people living on the shores of the North Atlantic. It is collected between late spring and early fall and allowed to dry in the sun; this drying treatment turns the plants very dark red to black in color. Traditionally, dulse has been eaten with dried fish, butter, and potatoes. It is also said to be rolled into a sort of chewing tobacco, a use for which its leathery texture seems well suited. It has been used as cattle food.

30. *Halosaccion glandiforme*

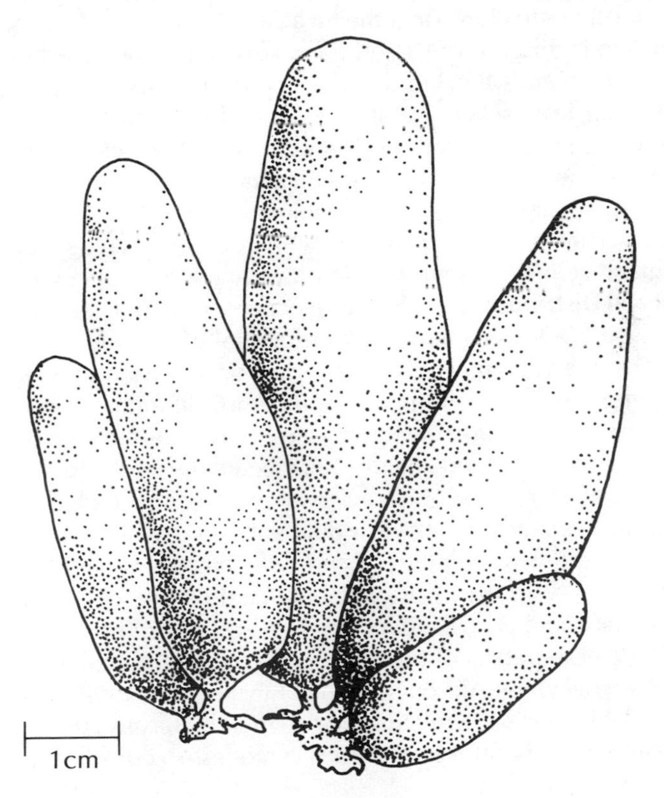

Rhodymenia pertusa (Postels and Ruprecht) J. Agardh *(Pl. 14)*

Description: A large, elongate (up to 1 m), red blade with numerous elliptical perforations, *R. pertusa* is somewhat thicker than its close relative, *Palmaria palmata,* and most of the plants are not branched. It is bright red in color and not as transparent as *P. palmata.* It has a small discoid holdfast and a very short stipe which may have several smaller bladelets arising from it.
Habitat and Distribution: Although *R. pertusa* is occasionally found in the lower intertidal zone, it occurs mostly in rocky subtidal regions. Most abundant in spring and early summer, it may be difficult to find at other times of the year. Sometimes it is cast ashore on beaches where seaweed drifts are abundant. It occurs from Oregon to Alaska.
Comments: In addition to the species of *Rhodymenia* described here, there are several smaller and less noticeable species found in Northwest waters.

Antithamnion defectum Kylin *(Pl. 13)*

Description: *Antithamnion* is a soft, small, branched filamentous alga. It is very beautiful and delicate when submerged in water, for then it appears bright red in color and fine, delicate and bushy in stature. When taken out of the water, it tends to clump together in a dark red, soggy lump, and its delicate construction is obscured. A good hand lens or low-powered microscope is essential to observe the branching pattern, a feature used to distinguish *Antithamnion* from several related genera which resemble it in overall appearance. The plants are typically only 2 to 4 cm high and are attached to the substrate by fine rhizoids which originate near the base of the plant and are generally quite difficult to find. The side branches of *Antithamnion* have an opposite arrangement. Many species of *Antithamnion* have special shiny cells called gland cells which are visible only with some optical aid.
Habitat and Distribution: *A. defectum* occurs commonly on a wide variety of substrates in the lower intertidal and subtidal zones and over a wide geographical range. You may also find it cast up in beach drift. It is relatively common from British Columbia to southern California.
Comments: *Antithamnion defectum* is just one representative of a number of small, closely related red algae, quite beautiful and fairly common, but requiring critical microscopic observation for correct identification. One common and related species in a different genus which you will certainly encounter is *Antithamnionella pacifica,* which is often found on old stipes of the kelp, *Nereocystis luetkeana.* Other related genera include *Hollenbergia, Scagelia, Platythamnion, Callithamnion, Pleonosporium,* and *Griffithsia,* all of which differ in details of branching patterns, reproductive structures, cell size and arrangement. Although many of these algae have remarkably large cells, the entire plants are relatively small in size. Because they can be grown under controlled laboratory conditions in modest-sized culture vessels, many

of these algae, especially *Griffithsia,* are superb organisms for experimental laboratory studies on plant growth and development and on how their development is controlled by both external environmental factors and internal intracellular control mechanisms.

Ceramium gardneri Kylin *(Pl. 13)*

Description: Most *Ceramium* plants are of small stature (6 cm or less) and are composed of branching filaments made up of fairly large, barrel-shaped cells. The cells are covered near their end walls with a band of much smaller cells. The degree of banding or cortication, as it is called, depends on the particular species. In *C. gardneri,* the bands are fairly narrow with the middle of the larger, older cells uncovered, so that this plant appears to have small horizontal stripes on it. The stripes are quite easily seen with the naked eye or with a hand lens. The very tips of the branches of *Ceramium* are composed of very small cells, and the branch tips often curve toward one another like pincers. The branching pattern is often very regular. The plants may range from pale to deep red in color.
Habitat and Distribution: *Ceramium* is often epiphytic on other algae in the lower intertidal and upper subtidal zones and may be found cast ashore in drift material. This species is found from northern Washington to central California; other species are common in many seas.
Comments: There are about half a dozen species of *Ceramium* found in the Northwest, and most require microscopic examination for identification. With its bands of corticating cells, *C. gardneri* is a good example of an intermediate degree of complexity in thallus organization. In less complex genera such as *Antithamnion,* the cells of the main axis are not usually corticated; in other more complex genera, such as *Microcladia, Delesseria,* and *Laurencia,* there is a main axis of fairly large cells, but these axes and branches are completely covered with layers of smaller cells.

Microcladia borealis Ruprecht *(Fig. 31)*

Description: *M. borealis* forms erect or prostrate tufts (to 18 cm high, usually shorter) which are attached to the substrate by an inconspicuous prostrate rhizoidal branch system. The plants are brownish red to very dark purple red in color. They have a regular and distinctive branching pattern in which all the side branches are located on one side of a main axis. Most branches lie in the same plane, and the plants appear flattened. The branch tips are quite small but do exhibit a pincerlike appearance.
Habitat and Distribution: *Microcladia borealis* is often abundant on rocks in the mid- to upper intertidal zone where they may occur individually or in fairly dense stands. This species is found from the Aleutian Islands to California.

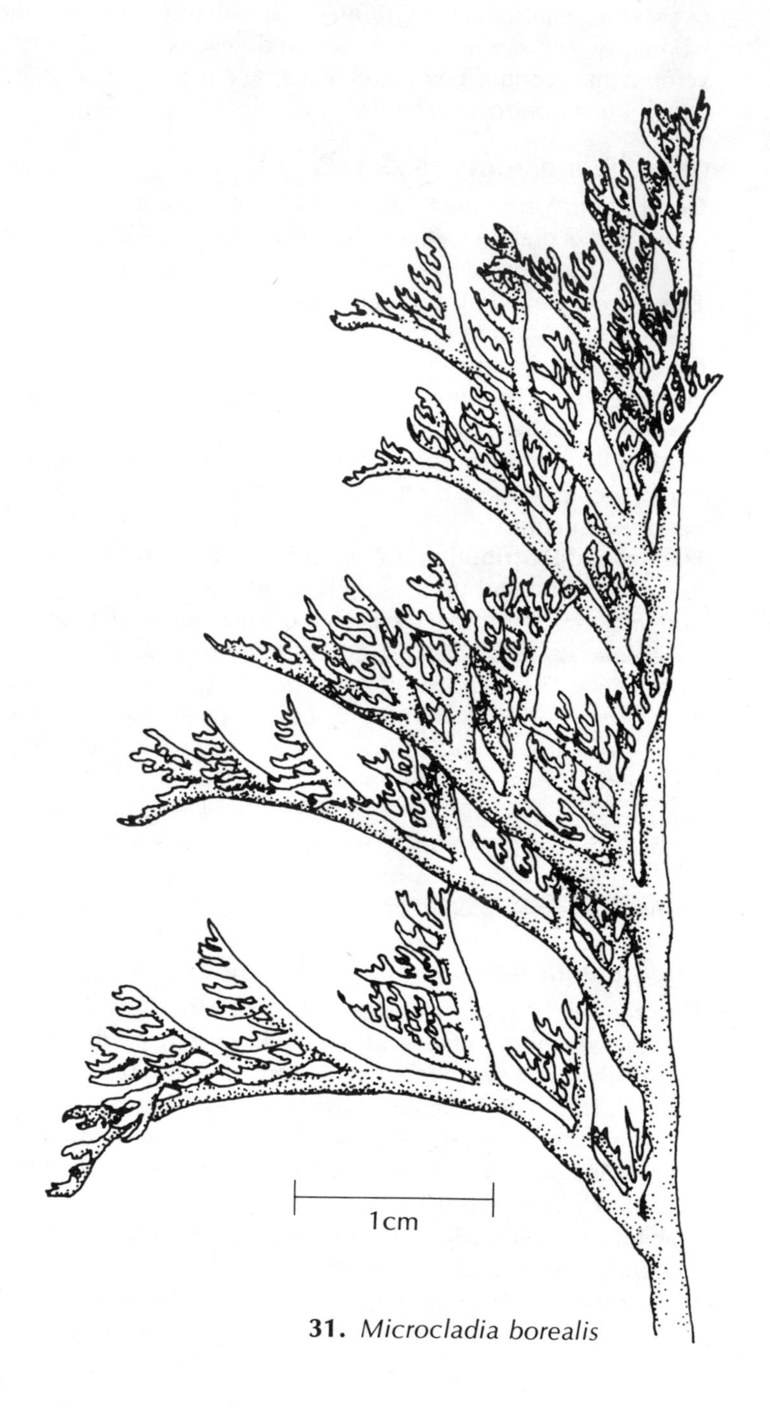

31. *Microcladia borealis*

Comments: Another common species of *Microcladia, M. coulteri,* is much larger (up to 35 cm long) and has branches on both sides of its main axis. Most of its branching is restricted to one plane; however, the tips of the branches appear bushier than *M. borealis. M. coulteri* is often found attached to other red algae, such as *Prionitis* and *Gigartina,* in the lower intertidal and subtidal zones.

Delesseria decipiens J. Agardh *(Pl. 14)*

Description: *Delesseria* is a strikingly beautiful seaweed. It has a small discoid holdfast, a conspicuous, often branched main axis with a prominent midrib, and delicate, flattened winglike projections 1 to 2 cm wide. Side branches bear small leaflets which arise at regular intervals from the midrib of the larger branches. *Delesseria* may be up to 35 cm tall and is usually deep, bright red in color. Dying plants which have been exposed to drying or somehow damaged may appear bright orange red in the damaged area. These plants have a fragile texture; the lateral wings of the branches and leaflets are in fact only one cell thick.
Habitat and Distribution: Occurring from Alaska to California, *Delesseria* is found in the lower intertidal and subtidal zones. It may also be found in fresh beach drifts.
Comments: This beautiful alga is fairly easy to find in the lower intertidal zone in early spring. Its rate of photosynthesis reaches a maximum at very low light intensities, and it seems to be damaged by direct, bright sunlight. This may explain why *Delesseria* is often found in shallow waters in late winter and early spring but becomes much rarer in such places when the days are brighter and the lowest tides occur during daylight hours.

Polyneura latissima (Harvey) Kylin *(Fig. 32)*

Description: Plants of this species consist of a large, long (10 to 30 cm), fairly broad (10 to 15 cm), and sometimes divided or lobed blade which is usually bright red in color. The texture is quite crinkly, and there is an easily noticeable network of slightly thickened "veins" throughout the blade. When fertile, the plants have small dark red, bumpy spots between the veins. The blade emerges from a small cylindrical stipe, and a small discoid holdfast anchors the plant to the substrate.
Habitat and Distribution: This species is found from Alaska to Mexico. It grows on rocks, worm tubes, and other substrates in the lowest part of the intertidal and subtidal zones to depths of 15 to 20 m.
Comments: *Polyneura,* a good example of a shade-loving seaweed, is often found on the underside or shady side of floating docks.

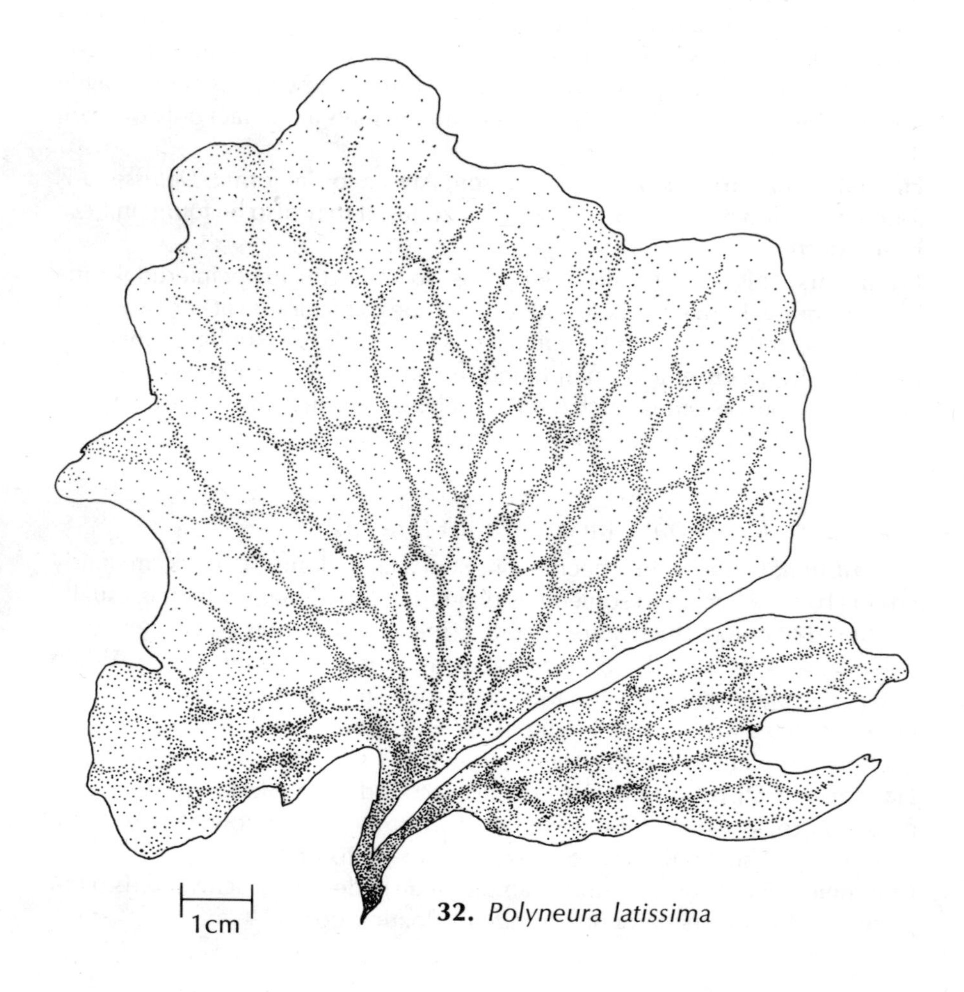

32. *Polyneura latissima*

Botryoglossum farlowianum (J. Agardh) DeToni *(Pl. 14)*

Description: *Botryoglossum* consists of a small, irregularly lobed holdfast from which a cylindrical stipe arises. Depending on the age and condition of the plant, there may be conspicuous "wings" of tissue on the stipe, or these may be worn off. The stipe branches frequently and then broadens into a 10 to 15 cm long, repeatedly branched, and divided blade. Each of the blade segments is 1.5 to 2 cm wide and 4 to 5 cm long. A faint network of veins radiates from the stipe and branches out into the blade segments. The margins of the blades are fringed with small outgrowths which give a ruffled look to the edge of the plant.

Habitat and Distribution: *B. farlowianum* occurs in the low intertidal and subtidal zones from British Columbia to Mexico.

Comments: Within the range of *B. farlowianum,* there is one closely related species, *B. ruprechtianum;* there are also at least two closely related genera, *Cryptopleura* and *Hymenena,* which superficially resemble *Botryoglossum.*

Polysiphonia pacifica Hollenberg *(Fig. 33)*

Description: *P. pacifica* forms tufts or plumes up to 15 cm in length. The plants are much branched, and a good hand lens or low-powered microscope is needed to discern structural details. Each branch (0.1 to 0.2 mm in diameter) consists of a central axial cell surrounded and covered by equally long cells called pericentral cells. *P. pacifica* has four pericentral cells. The plants are reddish brown to black in color and are anchored to the substrate by a system of prostrate branches. Spores resulting from sexual reproduction are produced on branchlets inside special urn-shaped structures. Spores on asexual plants are produced adjacent to axial cells but underneath the pericentral cells, so they appear to be embedded in the branch. Spore-bearing branches are often dark in color and have a swollen appearance.

Habitat and Distribution: *Polysiphonia pacifica* occurs from Alaska to California on rocks and other objects in the lower intertidal and subtidal zones.

Comments: There are many species of *Polysiphonia* on the Pacific Coast; all require careful microscopic examination for correct identification. The number of pericentral cells varies among species, and the arrangement of certain special branches and other features must be observed in order to identify species. Since it is a common genus in many parts of the world, *Polysiphonia* is often used as an example of a "typical" red alga. Its life history, illustrated in *figure 10,* has been known for a long time.

Laurencia spectabilis Postels and Ruprecht *(Fig. 34)*

Description: *Laurencia* plants are usually upright or draped stiffly over the rocks to which they are attached. They may be up to 20 cm tall, with a stiff,

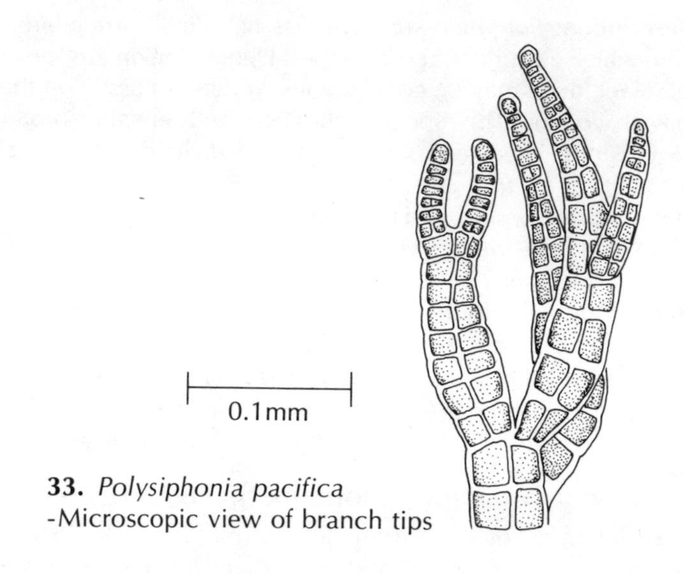

33. *Polysiphonia pacifica*
-Microscopic view of branch tips

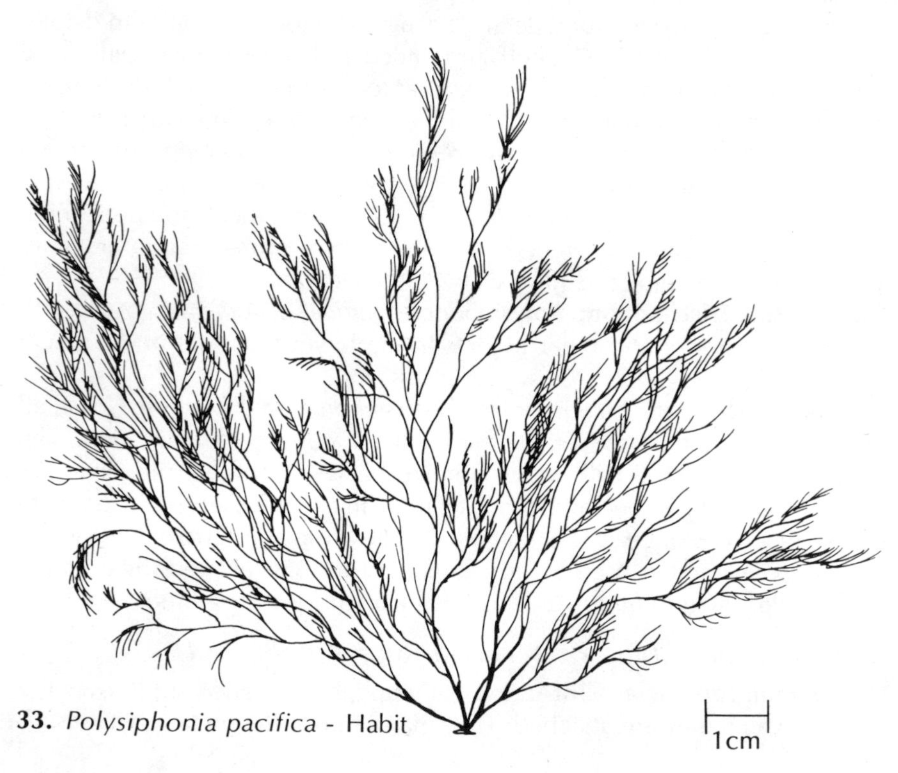

33. *Polysiphonia pacifica* - Habit

firm texture and branches typically 0.5 cm thick and 0.8 cm wide. Each main axis bears a series of short, blunt-tipped, somewhat flattened branches—a characteristic feature. *Laurencia* is usually deep purple red in color. It attaches to the substrate by a small discoid holdfast.
Habitat and Distribution: *Laurencia* occurs on rocks in the lower intertidal and upper subtidal regions from Alaska to California.
Comments: This species has been called pepper dulse, since it has a sharp, peppery taste. Try chewing a small piece to see if you like the flavor.

Rhodomela larix (Turner) C. Agardh *(Fig. 35)*

Description: This plant forms "ropy" strands up to 30 cm long and is usually dark reddish brown or black in color. It has a small discoid holdfast from which the main axis arises. The main axis may branch several times. All the branches are clothed with a dense covering of short (5 to 10 mm), radially arranged branchlets which nearly hide the major branches.
Habitat and Distribution: *R. larix* occurs on rocks in the mid- to lower intertidal zone. It is quite common on sheltered shorelines from Alaska to California.
Comments: In some places, *Rhodomela* forms a conspicuous part of the intertidal flora. It may be found in extensive bands where it is often associated with *Odonthalia* which resembles it in many ways. The saccate brown alga, *Soranthera,* is usually found growing only on *Rhodomela* in nature, even though *Soranthera* can be cultured without *Rhodomela* in the laboratory.

Odonthalia floccosa (Esper) Falkenberg *(Pl. 15)*

Description: This alga may be up to 40 cm long and is usually profusely branched. It is brownish to black in color with a small discoid holdfast. There are usually several conspicuous axes which are nearly cylindrical near the base and somewhat flattened near the tips. The major axes have a series of alternately arranged, repeatedly branching, smaller branchlets. Near the tips, the branchlets frequently appear tightly knotted.
Habitat and Distribution: From Alaska to California, *O. floccosa* is often common in the lower rocky intertidal zone.
Comments: There are three other species of *Odonthalia* in the Northwest. Since they intergrade somewhat, they can be difficult to identify.

Flowering Plants

The so-called sea grasses are not grasses at all in the botanical sense. They are flowering plants, closely related to certain pond weeds which live in fresh waters. You will certainly encounter these sea grasses in looking for algae, for many algae are found attached to the leaves of these aquatic flowering plants. The flowers of these plants are inconspicuous and vaguely resemble oddly shaped spikes of wheat. Worldwide, there are eight genera and many species of marine flowering plants. Only two of these genera occur in the Pacific Northwest waters.

Zostera marina Linnaeus

Description: *Zostera,* or eelgrass, has long (from 0.6 to 1 m to 3 to 4 m long, depending on the variety), flat, strap-shaped leaves more than 32 mm wide. The leaves are pale green; they arise in clusters from prostrate rhizomes which also produce roots.

Habitat and Distribution: *Zostera* forms "meadows" in quiet, protected bays with muddy bottoms. The shorter-leaved variety is usually found in the intertidal and upper subtidal zones, while the longer-leaved variety is always found in the subtidal region. Both varieties occur from Alaska to Mexico.

Comments: *Zostera* leaves have reportedly been used for insulation and mattress stuffing. Indians living along the Gulf of California harvested the seeds and ground them up as a source of flour which is reported to be bland

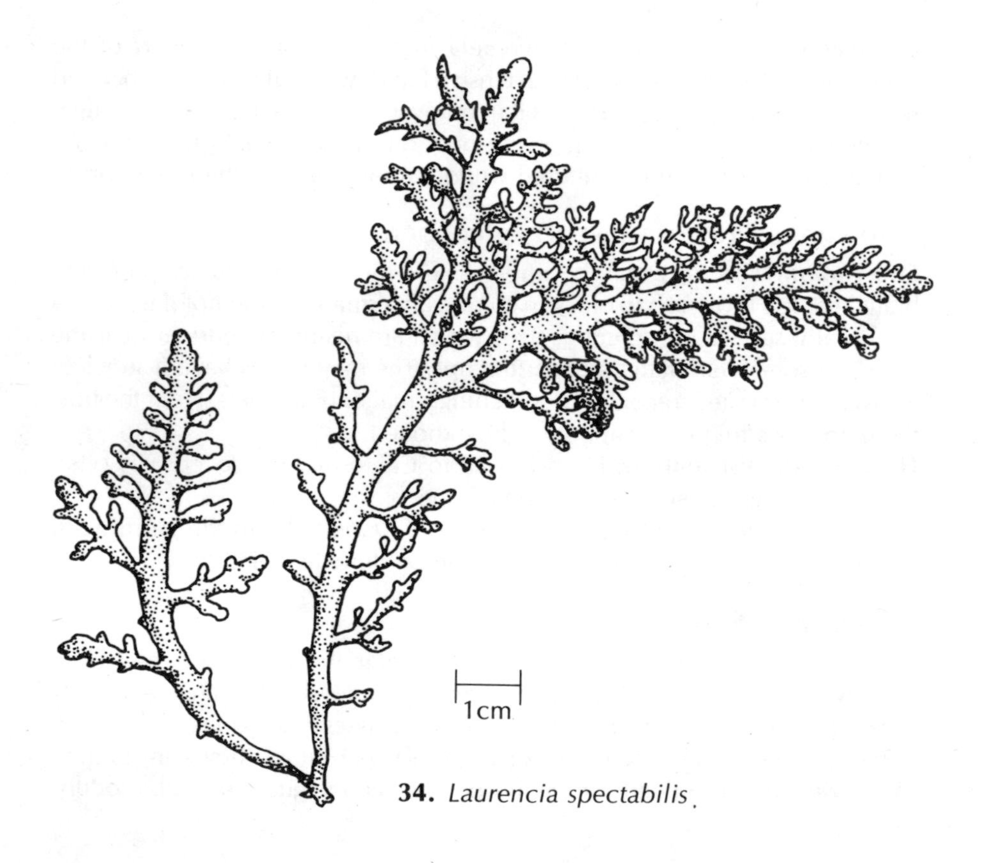

34. *Laurencia spectabilis*.

but just as nutritious as any terrestrially harvested grains.

Phyllospadix scouleri Hooker *(Pl. 15)*

Description: Like *Zostera, Phyllospadix* has long strap-shaped, bright green leaves (1 m long in *P. scouleri*) which are, however, narrower (20 to 32 mm wide) than its relative. The leaves arise from prostrate rhizomes which also bear roots.

Habitat and Distribution: *P. scouleri,* or surf grass, occurs in the lower intertidal and upper subtidal zones on exposed, rocky shores. It can be found from Alaska to Mexico.

Comments: A second species of *Phyllospadix, P. torreyi,* has thinner (less than 20 mm wide) and longer (to 3 m) leaves. It usually grows permanently submerged in rocky pools or in deeper water than *P. scouleri.* The small pink spots frequently seen on the leaves of *Phyllospadix* are the crustose corraline red alga, *Melobesia.*

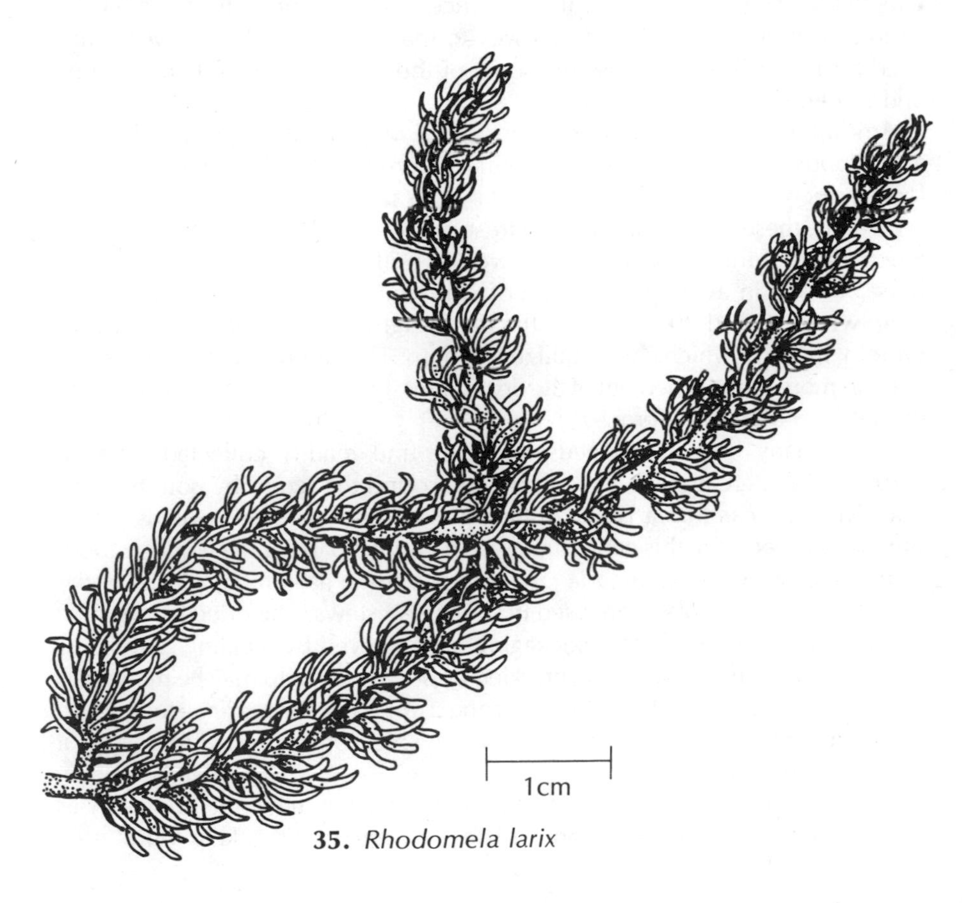

35. *Rhodomela larix*

Uses of Algae

V

Algae as Food

Algae have traditionally been used as food. Most algae are not particularly effective as sources of food calories, since much of the cell wall material cannot be digested by humans. However, many of the edible seaweeds are used for their distinct flavors, and some of them are rich in certain vitamins and minerals.

Among the green algae, the membranous or bladelike *Ulva* (sea lettuce) and *Monostroma,* as well as the tubular *Enteromorpha,* have been used as food, principally as condiments or flavoring in salads and soups. As with most seaweeds, these may be used when fresh, or they may be dried for later use. Since many of these seaweeds grow well in polluted waters, common sense must be used in avoiding ingestion of contaminated plants.

Between 10 and 20 percent of the dry weight of these seaweeds may be protein, much of which can be utilized by man. The carbohydrate component may be more than 50 percent of the dry weight, but often only a small fraction of that is in digestible form.

Since many brown seaweeds are large and readily collected, it is no surprise that a great many are consumed around the world by both humans and livestock. *Laminaria* is one of the better known edible seaweeds, and a number of species of this genus are used in various parts of the world as food. In the Orient, where *Laminaria japonica* is known as *kombu,* pieces of the dried and prepared blade are used in soups and stews. The dried blades may be shredded or rolled before packaging or preserved by pickling, salting, or even sugarcoating. Very finely shredded or powdered kelp may be made into small candies or cakes. In northern Europe, *Laminaria saccharina* (sometimes called sugar wrack) was used in similar ways; its stipes are said to possess a flavor reminiscent of peanuts.

A variety of other kelps, such as *Alaria,* are used in much the same way as *Laminaria.* One of the more popular uses of *Nereocystis*, a kelp found only along the Pacific Coast of North America, is as pickles, made from segments of the long, cylindrical stipe (see Chapter VIII for recipe). A few other species of brown algae, such as *Analipus* and even *Fucus,* have been eaten by coastal people.

Often, the flavor of one algal species is different from another. Some differences are due to the kinds and amounts of soluble sugars present, as well as the kinds and amounts of free amino acids. Many brown seaweeds have been used as fodder for sheep, cattle, and horses in coastal areas where abundant supplies of these seaweeds were readily available. The best results are obtained when the animal's diet is 3 to 10 percent seaweeds; however, some animals have been fed seaweeds exclusively for long periods of time. In contrast to man, these animals are able to digest the carbohydrates in seaweeds. In the tropics where most brown seaweeds are fairly small, the relatively large *Sargassum* has been used as pig feed.

Seaweeds are also used in food production as mulch and fertilizer. For this purpose, beach drift, which consists predominantly of brown seaweeds, is used. During the twelfth and thirteenth centuries, long before manuring was a common practice in most agriculture, one coastal farming region of northwest France was known as the golden belt, since beach drift seaweeds were readily available as manure. The productivity of these coastal farms was much higher than those farther inland where no such fertilizing was done. Seaweed mulch is used by present-day farmers in many coastal areas of the world. Seaweed manure tends to be rich in nitrogen and potash but low in phosphate; in regions of intensive agriculture, it is necessary to supplement seaweed manure with phosphate compounds.

Many red algae are used for food in a variety of dishes. Perhaps the most widely known is *Porphyra*, also called *nori* (mostly *P. tenera*) in Japan and *laver* (mostly *P. laciniata)* in the British Isles (see recipes in Chapter VIII). In Japan, a substantial industry is based on the cultivation of *nori.* Naturally occurring *nori* has been collected for centuries, the first *nori* farms being located around Tokyo Bay about 1700. In recent decades, *nori* cultivation has become quite sophisticated; many parts of the process are highly mechanized and quite closely controlled. Thousands of workers are employed in the raising, processing, and marketing of *nori*—a 250 million-dollar-a-year industry.

Nori is used mostly for flavoring in various soup and stew dishes, and it is also wrapped around small rolls of rice *(maki-sushi).* In the British Isles, *laver* has been used as a salad item; it is also fried in a great amount of fat as a breakfast dish. Dried, packaged *nori* is quite expensive. Usually, only small quantities are eaten, and it is quite nutritious. Typically, 25 percent of its dry weight is protein, much of which can be utilized by humans. It also contains vitamins, such as Vitamin C, and minerals, such as iodine, which is essential for normal thyroid function. Many of these compounds vary in concentration, depending on the season of the year and the condition of the plants. The indigestible bulk of many algae is said to prevent constipation; however, ingestion of excess quantities of some algae can lead to diarrhea. In the Northwest, a few of the many local species of *Porphyra* are said to taste like *nori,* and preliminary research is underway to test the feasibility of farming *nori* in this region.

Several other red algae which occur in the Northwest are edible. Often, these are species which have a greater number of devotees in other parts of the world—seaweeds, such as *Neoagardhiella, Gracilaria, Gracilariopsis, Palmaria,* and *Laurencia.* Many of these are used fresh, while others may be dried or pickled for later use. Some local seaweeds, such as *Iridaea* and *Gigartina,* can also be used to make dishes like blancmange (normally made from *Chondrus crispus* which does not occur here).

Industrial Polymers from Algae

These seaweeds are also important sources of marine polymers. While some parts of the algal food industry represent a substantial business enterprise (the *nori* industry, for example), in the United States, the seaweed polymer industry represents perhaps the largest seaweed-oriented industry. These marine polymers are obtained from two major algal groups: the brown algae (source of algin and related compounds), and the red algae (source of agar, carrageenan, and related products).

Seaweed polymers may be part of a finished product, as in ice cream, or they may be used during the processing of some product, such as beer. They are used in the manufacture of a tremendous variety of products from paint to pills.

Algin is composed of long molecules which are part of the cell walls of many brown algae, especially the large kelps and rockweeds. Along the coast of southern California, huge motorized barges mow and collect the upper parts of the fronds of the giant kelp, *Macrocystis.* The harvest is then taken to a factory where it is dried and processed to extract the algin. The extract—alginic acid—is typically converted to one of its salts, sodium or potassium alginate, before use. These salts dissolve readily in water, forming a thick, viscous solution. There are a number of methods for tailoring this viscosity to a particular purpose.

Algin is used in processing diverse products—textiles, charcoal briquets, paper, printer's inks, paints, car polishes, cosmetics, pharmaceuticals, candy bars, dairy products, bakery products, rubber, and many other items. While not commercially feasible on a large scale, methods have even been developed for making algin fibers or yarns for cloth.

In the late 1950s, a gradual decline in the extent of the kelp beds reached serious proportions, threatening to eliminate the raw materials of the California-based kelp industry. Thus, for over a decade, an extensive research project concerned with management and restoration of the naturally occurring kelp beds has been in progress. One of the first steps taken to restore the kelp beds was the control of populations of the numerous kelp-grazing sea urchins which were eating young kelp plants before they could reach a substantial size. The kelp restoration program seems to be successful. It has resulted in greatly increased knowledge of the biology of the giant kelp forests (see Chapter VI). In other parts of the world, such as Norway and Great Britain where other species of kelps are harvested, researchers are also investigating the factors controlling these valuable marine plants.

The major industrial polymers found in red algae are agar and carrageenan. Agar is obtained from seaweeds like *Gelidium* and *Gracilaria,* and there may be commercially harvestable quantities of these agar-weeds in southern California. Today, *Gelidium* and *Gracilaria* are harvested in Baja California. Although agar is processed in the United States, much of the production comes from Japan where raw material from Southeast Asia, Africa, Mexico, and South America is processed.

Agar is used principally in the making of microbiological culture media; it is also used in baking, confectionery, meat and poultry products, in desserts and beverages. Smaller amounts of agar are used in the pharmaceutical, dental, and other industries. Over a million pounds a year are used in the United States.

Carrageenan, obtained primarily from *Chondrus crispus,* a red seaweed native to New England and the Maritime Provinces is used in greater quantities than agar. Annual domestic production of refined carrageenan is over 10 million pounds. Demand for carrageenan has increased substantially over the years, and the raw material is now extracted from several seaweed species in temperate as well as tropical waters.

Carrageenan is used as a stabilizer and emulsifier in a wide variety of food and industrial products. By appropriate modification of the carrageenan molecule, it can often be tailored to a particular product. Increased worldwide demand has resulted in exploration and research into new and better ways of harvesting and managing natural stands of carrageenan-producing seaweeds for maximum yields consistent with judicious management of this living resource.

There also has been considerable stimulation of interest in the possibility of farming carrageenan-producing seaweeds by aquaculture. Studies of *Chondrus* are centered in Nova Scotia and New England. One tropical species, *Euchecuma,* is now being raised on small farms in the Philippines. In the Northwest, research on the feasibility of aquaculture of *Iridaea* and *Gigartina* has been underway in Washington since 1971, and related studies are also being conducted in British Columbia.

Aquaculture of Seaweeds

For centuries, users of seaweeds have relied on hunting and gathering methods of obtaining edible or industrially useful seaweeds. Much of the world's supply is still obtained this way. However, farming methods have been developed for the edible seaweed, *Porphyra (nori),* in Japan, for the edible *Laminaria (kombu)* in Japan and China, and research into methods for farming carrageenan-producing seaweeds is underway in several parts of the world. In the Pacific Northwest, these efforts presently are focused on *Iridaea* and *Gigartina.*

Nori cultivation began in Tokyo Bay around 1700. The basic technique involved placing additional substrates at depths where *Porphyra* normally grew. The substrate was composed of bundles of brushwood called *"hibi."* These bundles were firmly implanted in the mud during a low tide in such a way that the tops of the brush would be covered by high tides. "Planting" of *hibi* occurred in late fall at a time when *Porphyra* was observed to begin growing on natural substrates. If the *hibi* were planted at the proper time, *Porphyra* soon was growing on the brush, and plants could be harvested from January through March. Today, the cumbersome brushwood substrates have been replaced by systems of nets which are rigged so that they will be out of the water for four to four and one-half hours when the tide is low.

In the early 1950s, a British botanist studying the life history of *Porphyra* made a critical discovery which made it possible to more carefully control *Porphyra's* colonization of substrates. Professor K. M. Drew-Baker found that *Porphyra* has two different-looking phases in its life history: one of these is the familiar bladelike form; the other is small, filamentous, and inconspicuous. This filamentous phase had been named *Conchocelis* by earlier botanists who were unaware that it was actually the plant that produced the spores that colonized the *hibi* in late fall. Now, stocks of *Conchocelis* are grown in laboratory culture. When they begin to produce spores, the *hibi* nets are "seeded" with spores either in the lab or in the field—a much more reliable practice than that of natural seeding. After the seeded nets are set out, they are grown for awhile; then the nets with the young plants may be dried and stored in a deep freeze. When the most favorable growth period occurs, the nets that have been stored in the freezer are thawed and rigged again in the sea.

As demand for *nori* increased from about 1000 tons per year in 1910 to 15,000 tons per year in 1960, all the available tidelands were used and new methods for off-shore, deepwater cultivation were developed. Instead of being rigged from poles, the nets are suspended from anchored buoys. Thus, the area available for *nori* cultivation has been greatly increased. Many sophisticated and mechanized techniques are now employed in all phases of *nori* culture from seeding to harvesting.

Because of the high price and steady increase in demand for *nori*, preliminary studies are underway in the Northwest to determine if any of the local species of *Porphyra* have the proper taste, texture, and appearance to warrant cultivation. Should such a species be identified, research into the best way to raise large quantities would begin. One would expect initial efforts along such lines to rely heavily on methods developed and used in cultivation of Japanese species of *Porphyra*.

Increasing demand for the marine polymer, carrageenan, has led to the search for new sources, such as *Iridaea cordata* and *Gigartina exasperata* in the Pacific Northwest. Because the natural supply of these seaweeds appears to be limited by the amount of appropriately located substrate, farming on artificial substrates is under investigation. No traditional methods exist for aquaculture of these species, so research had to start with very basic problems, such as developing methods for growing and transplanting them in field experiments and for raising them in the laboratory.

In Washington, researchers supported by the University of Washington Sea Grant Program, the Department of Natural Resources, the National Marine Fisheries Service, and two industrial processors of marine polymers (Marine Colloids, Inc. and General Mills Chemicals, Inc.) have cooperated in various phases of this project. Thus far, the feasibility of transplanting and growing *Iridaea* and *Gigartina* on artificial substrates (nets, plastic rope, concrete blocks) has been demonstrated. The optimum depth for growth in the field and the optimum temperature and light intensity for laboratory cultivation have been determined.

Present work is centered on developing a means of large scale colonization of artificial substrates (nets) by spores from these plants. Both field and laboratory methods have been successfully tested on a small scale. Plants growing on artificial substrates have been observed to double their weight in less than two weeks during the optimum growing season between late spring and summer. Once plants have begun growing on artificial substrates, there is reason to hope that a long-term, sustained yield can be obtained from these substrates, since the base of a plant of *I. cordata* or *G. exasperata,* the encrusting holdfast, is perennial, while the large blade which is harvested is produced annually.

Another method of growing these seaweeds, using large vats through which seawater is pumped, is under investigation. Vat cultivation is advantageous in that it can be carried out on a more or less continuous basis, while the material cultivated on aritificial substrates must be harvested in batches.

Thus, it appears that seaweed aquaculture may join salmon and shellfish aquaculture as a marine-oriented industry in the sheltered inland waters of the Northwest.

Interactions of Algae and Animals

The spectrum of interactions among organisms covers a broad range from little or none, to very positive impacts in which both organisms benefit from the interaction, to very negative ones, including such cases as parasitism or toxic interactions. Let us consider briefly three examples of interactions between algae and animals: symbiosis, grazing, and toxic interactions.

Symbiosis

Symbiosis is the living together or association of two dissimilar organisms. Often, it implies an association of mutual benefit to the two organisms; sometimes only one of the organisms seems to benefit, while the other suffers no obvious ill effects from the association.

There are many symbiotic associations between local algae and animals. Most of the algal partners in such associations are microscopic, such as the association which occurs between microscopic algae and the large green sea anemone, *Anthopleura xanthogrammica (pl. 16)*. This anemone, measuring up to 20 cm in diameter, occurs from Alaska to Panama and generally lives in the lowest depths of the intertidal zone or in permanently submerged tide pools. Its vivid green color is due to the symbiotic green and golden algae in its tissues. If they are green in color, such algae are called *zoochlorellae;* if they are golden brown in color they are called *zooxanthellae*. Such symbiotic algae are found in the tissues of many marine organisms, including corals, giant clams, sponges, and worms. Many of the animals with symbiotic algae orient their bodies so that the algae receive adequate sunlight for photosynthesis. In some of the more thoroughly studied cases, it has been shown that photosynthetically produced food molecules from the algae are utilized by the animal. This is undoubtedly beneficial to the animal. On the other hand, some of the waste products from the animal can be used by the algae, and the algae have a moderately protected place in which to live.

Grazing

Almost all plant life serves as food for some type of animal life. Since they are plants, seaweeds are no exception to this general rule. Some algae are quite obviously grazed on by animals.

Perhaps the best example of such grazing is that of sea urchins *(Strongylocentrotus* spp.) grazing on kelp. Feeding experiments have shown that sea urchins have a preference for the blades of algae such as *Nereocystis,* while they will only eat *Agarum* as a last resort. Divers have reported migrations of thousands of sea urchins into kelp beds, an invasion which sometimes results in the area being stripped of most algal growth. Sometimes the sturdiest part of the stipe or holdfast of the larger kelps remains. Usually the rock-hard coralline algae are little affected, but most of the more palatable seaweeds are eaten.

Little is known about the migrations, invasions, or population explosions of urchins in Northwest waters, but a somewhat better understanding of the interactions between grazing sea urchins and the southern California forests of the giant kelp, *Macrocystis,* has been found. By the late 1950s, a steady decline in the valuable kelp forests threatened the source of raw materials of the algin industry. Three possible causes of this decline were suspected: pollution which could slow the kelp growth rate by increasing water turbidity (cloudiness) or by adding harmful substances to the water; a warming trend which might be deleterious to the kelps; and finally, the near extermination of sea otters, natural predators of the kelp-eating urchins. After several years of study, it was concluded that urchin control (once provided by the urchin-eating sea otters) was the key to reestablishing the kelp beds. It was found that quicklime, dumped on the marauding populations of sea urchins, could be used to selectively destroy the urchins. Once the urchin population was controlled, the kelps were able to reach a critical size and productivity. Then the kelps grew fast enough, so that urchins fed on the detritus from the kelp forest, rather than on the healthy young kelp plants. In addition, strains of *Macrocystis* favoring warmer temperatures were transplanted into some areas. In recent years, the southern California kelp forests have been restored substantially. They now cover almost as much area as they did in the early part of this century.

Red Tides

During the warmer months of the year, coastal waters may suddenly turn the color of tomato soup—a phenomenon appropriately called red tide. Such water conditions are usually due to the rapid, explosive growth of a single species of microscopic planktonic alga.

In Washington waters, the only species of alga which colors the water red is a large-celled dinoflagellate called *Noctiluca*. Other algal species may cause the water to turn other colors—pink, orange, amber, or brownish red—but the phenomenon is usually still called a red tide. Some red tides are harmless, but often the organisms which undergo this explosive growth produce a deadly toxin which causes paralytic shellfish poisoning when ingested by humans.

In Northwest waters, the dinoflagellate, *Gonyaulax catenella (pl. 16),* is associated with deadly red tides. When it forms dense "blooms" of millions upon millions of cells, it may color the water rusty red. However, it can be abundant enough to cause shellfish poisoning without being dense enough to cause the water to be obviously colored. *G. catenella* and other red tide organisms increase when particular water conditions and particular concentrations of nutrients (which have not yet been adequately defined) occur. Filter-feeding shellfish, such as oysters, clams, and mussels, strain these algae from the water as part of their food. They themselves do not succumb to the toxin, but they do store and concentrate the poison when they eat millions of cells of the red tide organism. Shellfish may also retain the toxin long after the bloom of red tide organisms is over.

If a human ingests even a small amount of toxin-containing shellfish, the powerful nerve toxin may cause temporary paralysis and even result in death if breathing is impaired. For this reason, health and game department officials monitor the toxin level in shellfish. They have established seasons for taking shellfish in certain areas; areas are closed to shellfish collecting when the toxin concentration reaches dangerous levels. Frequently, red tide organisms are bioluminescent, producing beautiful displays of light when disturbed or jostled. It is reported that coastal Indians avoided eating shellfish during periods when the water was bioluminescent.

Seaweed Conservation

The popularity of marine biology has spelled disaster for many prime, accessible intertidal areas. In most places, the more colorful and conspicuous animals have been collected or uprooted from their habitats. Frequently, there is a general trample effect caused by multitudes of interested visitors who would not otherwise deliberately maul or loot the intertidal zone. Seaweeds may also be disturbed or uprooted by such trampling. As a consequence of such intense public interest and curiosity, most state or provincial conservation agencies have found it necessary to establish rules which protect marine life, by regulating the taking or possession of all sorts of marine and terrestrial organisms. Many marine parks and preserves have been established to conserve marine habitats and provide a refuge for such forms of life. In an effort to avoid disturbance due to curious or malicious intruders, marine botanists and zoologists quite often have had to select remote, nearly inaccessible and often treacherous stretches of coastline on which to conduct their field experiments.

Since much shoreland and even tideland is in private ownership, public parks often provide the best and only access to intertidal habitats. It is best just to examine and appreciate seaweeds and other marine organisms in their natural surroundings. If you must collect and preserve seaweeds, the proper procedures are outlined below. For eating purposes, it is important that seaweeds be healthy and fresh. Some are best used fresh, while others can be preserved by rapid drying (in sunlight or with gentle dry heat) or by pickling or salting. Fresh specimens of many seaweeds can be maintained for as long as twenty-four hours by keeping them cool, damp, and uncrowded.

For many years it has been a favorite pastime of seaweed fanciers to prepare pressed and dried seaweed specimens. Such collections serve several purposes: they provide a permanent record of a collection or observation; they are a useful reference for later comparison or identification; and they are often quite beautiful and aesthetically pleasing. Such dried specimens are prepared by spreading a wet seaweed specimen of appropriate size on a damp piece of sturdy high quality "botany" or "biology" drawing paper or high rag content "herbarium" paper. The seaweed is then covered by waxed paper, cheesecloth, or muslin. This combination is then sandwiched between large blotters or newspapers which absorb moisture. Finally, a piece of corrugated cardboard is placed external to each blotter to provide ventilation and hasten drying. A number of these sandwiches are then piled up, and the bundle secured by means of a board and heavy weight or a standard plant press frame and straps to keep the specimens flat. The cloth overlay, blotters, and corrugates must be changed daily until the specimens are dry. If drying is hastened by excess heat, the specimens may become too crisp and brittle. Most seaweeds will adhere to the paper by means of their own gums or mucilages. To be scientifically valuable, there should be only one species per sheet (unless you have collected an epiphyte or parasite), and the sheet should be labelled with the name of the seaweed, the date and place of collection, the name of the collector and any pertinent field notes. A serious collector should also maintain a notebook and number each specimen so that it can be cross-referenced in the notebook. The notebook entry should contain the same sort of information as the label on the specimen sheet. If a number of specimens are collected at one site, the notebook is the best place to record extensive additional information about the particular habitat or site. If you feel a compulsion to collect seaweeds, be prudent and collect only what you will use. Be sure to comply with all applicable regulations.

Seaweed Recipes

A few seaweeds are eaten "as is," while most are dried and added to soups, stews, and casseroles for flavoring or to provide a certain texture or color. In addition, palates differ considerably in their preferences, so some experimentation and/or modification of a basic recipe may improve it. For those desiring to engage in such experimentation, the following recipes are offered. Additional recipes may be found in Oriental and natural food cookbooks.

Nereocystis pickles

Gather approximately one large, healthy *Nereocystis* plant per pint of pickles. Cut the cylindrical stipe into sections 6 to 8 inches long. Place the sections in a large kettle and bring to a boil for 5 minutes. Drain and rinse the sections; then slice into thinner slices as for cucumber pickles. Boil slices for one hour until tender.

For the pickling juice, mix up the following ingredients (enough for 10 pints of pickles):

VINEGAR *2 pints*
WATER *1 pint*
LARGE ONIONS *3, chopped*
SUGAR *3 pounds*
LEMONS *3, sliced*
CINNAMON BARK STICKS 6
WHOLE CLOVES *2 tablespoons*
MACE *a pinch*

Add pickling mixture to the drained kelp slices. Cook for half an hour, then seal in pint jars. You may also wish to experiment with other pickle recipes.

Kombu recipes

For authentic *kombu,* it is best to purchase some at an Oriental food store; however, young and tender specimens of local *Laminaria saccharina, Hedophyllum sessile,* or *Alaria marginata* serve as good substitutes in many *kombu* recipes.

Kombu Maki with Pork Center *(Seaweed Roll)*

DRIED KOMBU *8 ounces*
PORK *1 pound*
SOY SAUCE *⅓ cup*
SALT *1 teaspoon*
SUGAR *3 tablespoons*
MONOSODIUM GLUTAMATE *½ teaspoon*

Wash *kombu* well and cut into strips 2½ by 5 inches. Cut pork into ½-inch cubes. Place a piece of pork on one end of each *kombu* strip, roll and secure with string. Put the assembled rolls into a saucepan, add water to cover and cook until tender (3 hours; only 1 hour in a pressure cooker). Add remaining seasonings and cook for 30 minutes.

Kombu Soup Stock

Boil 3 cups water and several 4-inch pieces of *kombu* together for 3 minutes; then remove *kombu* and use soup stock for sukiyaki, etc.

Small pieces of *kombu* may be added to other soups while cooking.

Nori recipes

For these, try using Japanese *nori* first, then experiment with local species of *Porphyra* for comparison.

Nori soup

The simplest way to sample *nori* is by adding it to a commercially available chicken soup. To start from scratch, however, try the following recipe:

CHICKEN STOCK *4 cups*
NORI *2 sheets*
EGGS *2*
SALT and PEPPER *to taste*
SESAME OIL (optional) *½ teaspoon*
GREEN ONION STALKS *1 or 2, chopped*

Boil chicken stock; then add *nori* and stir with fork to loosen. When *nori* is soft, stir in well-beaten eggs and boil for a few seconds. Add salt and pepper. Before serving, add sesame oil; garnish with onion. Note: To make your own chicken stock, add 1 teaspoon of Monosodium Glutamate and 4 cubes of chicken or beef bouillon to 4 cups of water.

Toasted Kim *(Laver)*

NORI *6 sheets*
SALT *½ teaspoon*
SESAME OIL *1 ½ tablespoons*

To clean, rub each sheet of *nori*. Mix the salt and oil and rub a thin coating of this mixture on one side of each sheet. Lay the sheets on top of one another; roll them up and let stand 5 minutes. Unroll and cook each sheet in a hot pan over low heat until crisp. Cut each sheet into four pieces and serve with hot rice. Although actually a Korean recipe, this resembles one manner in which *laver* was prepared in the British Isles.

Blancmange

While this pudding recipe was really developed for use with *Chondrus crispus* (Irish moss), it will work with most of the carrageenan-containing seaweeds. Local species of *Iridaea* and *Gigartina,* which typically contain about 50 percent carrageenan (percent dry weight), may be used as substitutes for *Chondrus*.

Thoroughly wash the seaweed in several changes of freshwater. Cut into small pieces; then put 1 cup of seaweed into a cheesecloth bag and tie the bag securely. Put the bag into 1 quart of milk in the top of a double boiler and cook over boiling water for 30 minutes. Stir occasionally and press the bag against the side of the pan with a spoon to extract as much carrageenan as possible. If the milk is not stirred, the extract will tend to gel around the bag. After 30 minutes, remove and discard the bag. Add 1 to 1 ½ cups sugar and a pinch of salt to the milk and allow to partly cool. When it begins to cool, add any fruit to flavor it, pour into molds and chill to set the gel. Aspic recipes may also be made using seaweed gels instead of gelatin by heating the seaweed in vegetable juice instead of in milk.

Selected References

The following list of references provides a guide to more detailed treatments of local seaweeds, reference to some books which contain further information on the biology and uses of seaweeds, and further references to the general subject of marine biology especially with reference to the Pacific Coast.

Floristic Treatments

Abbot, I. A. and Hollenberg, G. J. 1976. *Marine Algae of California.* Stanford: Stanford University Press. Many species of algae which occur in California are also found in Oregon, Washington, and British Columbia; this useful reference has numerous line drawings, descriptions, and keys to genera and species.

Scagel, Robert F. 1967. *Guide to Common Seaweeds of British Columbia.* Handbook no. 27. Victoria, B. C.: British Columbia Provincial Museum, Department of Recreation and Conservation. This book has keys to the genera of most of the commonly encountered seaweeds of the Pacific Northwest. It is profusely illustrated with line drawings. An updated revision is in preparation.

For all the species of marine algae known from the waters of Washington and British Columbia, consult the following:

Scagel, Robert F. 1966. *Marine Algae of British Columbia and Northern Washington.* Part I: Chlorophyceae (Green Algae). Ottawa: National Museum of Canada. Bulletin no. 207.

Widdowson, Thomas B. 1973. "The marine algae of British Columbia and Northern Washington: revised list and keys. Part I. Phaeophyceae (brown algae)." *Syesis,* vol. 6, pp. 81-96.

Widdowson, Thomas B. 1974. "The marine algae of British Columbia and Northern Washington: revised list and keys. Part II. Rhodophyceae (red algae)." *Syesis,* vol. 7, pp. 143-186.

Reference Books

Boney, Arthur D. 1966. *A Biology of Marine Algae*. London: Hutchinson.

Chapman, Valentine J. 1970. *Seaweeds and Their Uses*. 2d ed. New York: Halsted. This book is a compendium of ancient and modern uses of algae all over the world.

Dawson, Elmer Y. 1966. *Marine Botany: An Introduction*. New York: Holt, Rinehart and Winston. Both of the above are widely used texts which contain a wealth of information about marine algae.

Marine Biology

Kozloff, Eugene N. 1973. *Seashore Life of Puget Sound, the Strait of Georgia, and the San Juan Archipelago*. Seattle: University of Washington Press. This copiously illustrated book is a comprehensive guide to the marine life of the inland waters of Washington. It emphasizes the commonly encountered species of animals and plants and is organized around the types of habitats encountered in the area.

Ricketts, Edward F. and Calvin, Jack. 1968. *Between Pacific Tides*. 4th ed. Stanford: Stanford University Press. This is an interesting and very readable book dealing with seashore life on the Pacific Coast. It too is organized around habitats and is copiously illustrated and has an excellent bibliography.

Stephenson, T. A. and Stephenson, Anne. 1972. *Life Between Tidemarks on Rocky Shores*. San Francisco: W. H. Freeman. A compendium of their life's study of the intertidal life in many parts of the world.

Index

Bold face numerals refer to pages on which black-and-white figures are located.

Other Paperbacks from Pacific Search Press

Cooking

Bone Appétit! Natural Foods for Pets by Frances Sheridan Goulart. Treat your pet to some home-cooked meals made only with pure, natural ingredients. Recipes fit for both man and beast! Drawings. 96 pp. $2.95.

The Carrot Cookbook by Ann Saling. Over 200 mouth-watering recipes. 160 pp. $3.50.

The Dogfish Cookbook by Russ Mohney. Over 65 piscine delights. Cartoons and drawings. 108 pp. $1.95.

The Green Tomato Cookbook by Paula Simmons. More than 80 solutions to the bumper crop. 96 pp. $2.95.

Why Wild Edibles? The Joys of Finding, Fixing, and Tasting–West of the Rockies by Russ Mohney. Color and black-and-white photos plus illustrations. 320 pp. $6.95.

Wild Mushroom Recipes by the Puget Sound Mycological Society. 2d edition. Over 200 recipes. 176 pp. $6.95.

The Zucchini Cookbook by Paula Simmons. Revised and enlarged 2d edition. Over 150 tasty creations. 160 pp. $3.50.

Nature

Butterflies Afield in the Pacific Northwest by William Neill/Douglas Hepburn, photography. Lovely guide with 74 unusual color photos of living butterflies. 96 pp. $5.95.

Cascade Companion by Susan Schwartz/Bob and Ira Spring, photography. Nature and history of the Washington Cascades. Black-and-white photos, maps. 160 pp. $5.95.

Little Mammals of the Pacific Northwest by Ellen Kritzman. The only book of its kind devoted solely to the Northwest's little mammals. 48 color and black-and-white photos, distribution map, index. 128 pp. $5.95.

Fire and Ice: The Cascade Volcanoes by Stephen L. Harris. Copublished with The Mountaineers. Black-and-white photos and drawings, maps. 320 pp. $7.50.

Living Shores of the Pacific Northwest by Lynwood Smith/Bernard Nist, photography. Fascinating guide to seashore life. Over 140 photos, 110 in color. 160 pp. $9.95.

by Minnie Rose Lov-

hine Haley. Extraor-

128 pp. $5.50.

Vaters compiled by